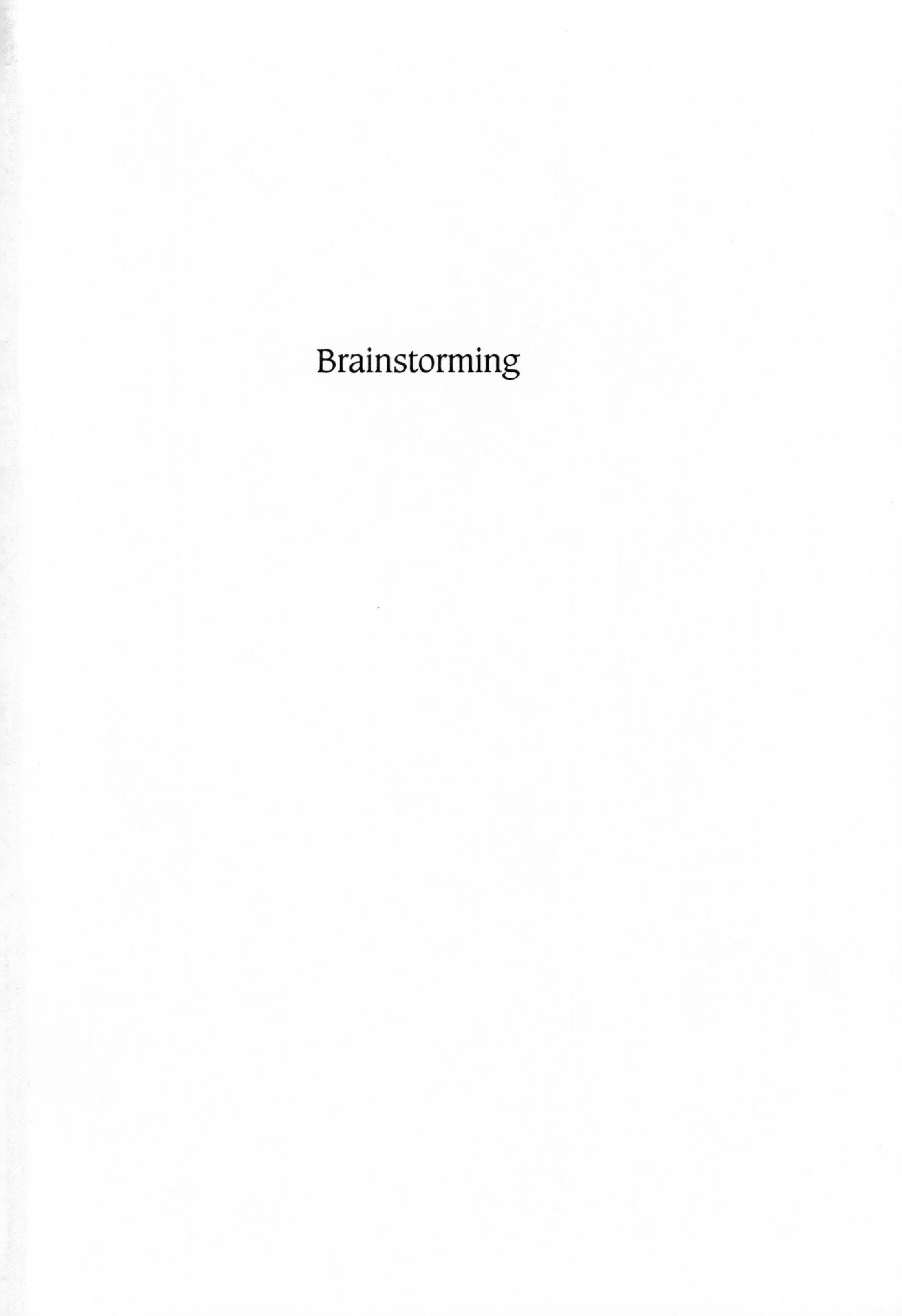

Brainstorming

Brainstorming

The Science and Politics of Opiate Research

•

SOLOMON H. SNYDER

HARVARD UNIVERSITY PRESS
Cambridge, Massachusetts
London, England
1989

Printed in the United States of America
10 9 8 7 6 5 4 3 2 1

This book is printed on acid-free paper, and its binding materials have been chosen for strength and durability.

Library of Congress Cataloging-in-Publication Data

Snyder, Solomon H., 1938–
Brainstorming : the science and politics of opiate research / Solomon H. Snyder.
p. cm.
Includes index.
ISBN 0-674-08048-3
1. Endorphins—Receptors—Research. 2. Endorphins—Receptors—Research—Political aspects—United States.
I. Title.
QP552.E53S68 1989 89-11096
612.8′22–dc20 CIP

Designed by Gwen Frankfeldt

To Julius Axelrod, an unsurpassed teacher,
and to all my students who made this work possible

Contents

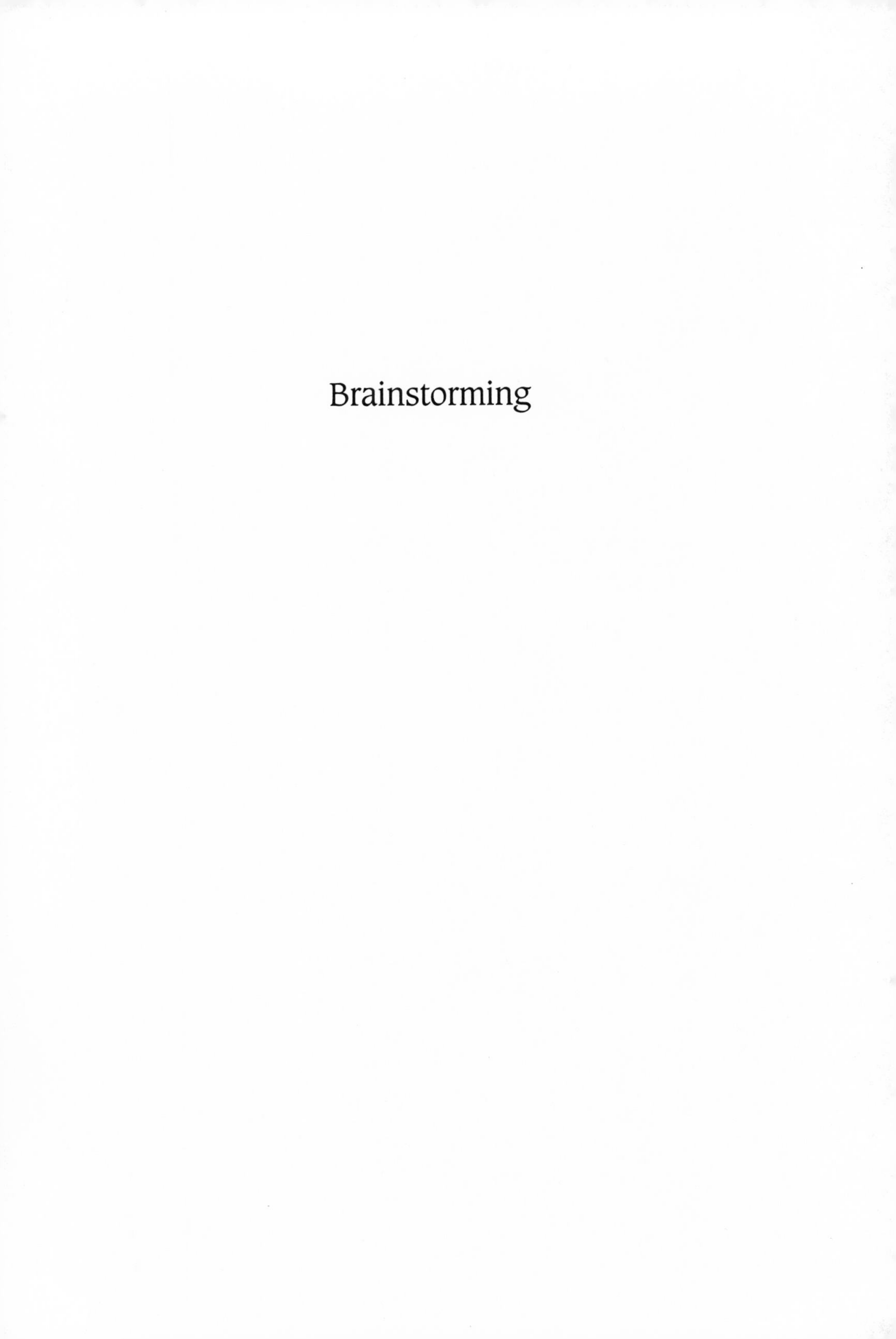

Brainstorming

Prologue

One of the founders of The Johns Hopkins School of Medicine in Baltimore, where I teach and run a neuroscience laboratory, was Sir William Osler, a leading physician at the turn of this century. Like my own mentor, the Nobel laureate Julius Axelrod, Osler often repeated axioms to his students—choice bits of wisdom reflecting a lifetime of experience. One of Sir William's axioms went something like this: "The physician who knows syphilis well knows all of medicine." In Osler's day, before screening for early detection was routine and antibiotics were widely available, syphilis was a common and deadly disease. Over the course of its three stages, it impaired every organ system—from circulation and the skin to reproduction. In its final phase, after the syphilitic organism entered the brain, it destroyed the mind. Understanding how this one disease disrupts function, Osler believed, would enable his young physicians to understand both the normal and abnormal physiology of the entire body. Equally important, it would help them appreciate the social and psychiatric dimensions of a devastating illness.

Just as Osler asked his students to master one disease, he also asked them to master a single drug—opium. Sir William chose opium, as he had chosen syphilis, for its incredible range of effects throughout the body. Opium was one of the most widely used medications in the nineteenth century. Its most obvious application—to relieve pain—had been known since time immemorial. Other well-established and effective uses in Osler's time included treatment of congestive heart failure, life-threatening dysentery as well as more benign diarrhea, and emotional disorders such as depression, anxiety, and insomnia. A standard British textbook of therapeutics, written by one Dr. Pereira in 1838 and popular in medical schools in the United States, recommended opiate extracts "to mitigate pain, to allay spasm, to promote sleep, to relieve nervous restlessness, to reduce perspiration and check profuse discharges from the bronchial tubes and gastrointestinal canal."

Today, opiates—principally morphine—are prescribed almost exclusively for short-term relief from pain. Physicians do not employ them routinely for other purposes because opiates are highly addictive, and because reasonably effective nonaddictive substitutes are now available. But if opiates could be designed that retained all the beneficial therapeutic effects while removing even the faintest hint of addictive potential, one might anticipate uses as widespread as in the nineteenth century. Nonaddicting opiates would be far superior to the synthetic sleeping pills, anti-anxiety medications, antidepressants, even cough medicines now in use. Many patients who today suffer almost unbearable chronic pain—from back problems, peripheral nerve damage, cancer—could find relief without fear of addiction to drugs that require ever larger doses to be effective. A nonaddicting opiate would also bring hope to the thousands of heroin addicts seeking a reprieve from their devastating disease. And once the underlying biological causes of opiate dependence were better understood, perhaps drugs to treat other addictions, including alcohol, might also be designed.

The most crucial step in learning how to create nonaddictive opiates is to understand the biochemical events in the brain that cause addiction. In 1972, with funds appropriated by Congress in response to President Nixon's call for a "war on drugs," my colleagues and I conducted a number of studies that pinpointed the molecular sites at which opiates act in the brain—the opiate receptors. My hope was that a new understanding of how opiates work would clarify the causes of addiction and lead to new approaches to this tragic and costly disease.

That hope is yet to be realized. Nonaddicting opiates have not yet been developed; and as the AIDS epidemic grows, especially among intravenous drug users, their lovers, and children, the problem of drug abuse weighs on the minds of citizens and politicians today much as it did in the early 1970s. But other medical and scientific benefits have accrued from receptor research that I could not imagine when I first became involved. Neuroscientists know not only where opiate receptors are found in the brain but also why they are there—to serve as binding sites for the body's own endogenous morphine, which regulates pain, mood, and a host of other physiological functions.

This book is the story of both the scientific and the political events that led to the discovery of opiate receptors, and of subsequent advances in receptor research that have revolutionized our understanding of the brain. The two worlds of science and politics are not easily separated. Science becomes embroiled in politics at every level of organization, from the Oval Office to the lab bench. As James Watson's *Double Helix* made clear, the politics of science feeds upon the same emotional frailties and uncertainties that plague other human enterprises. In making research choices that impinge on their careers, scientists do not always dispassionately plot out the most direct path to scientific progress. A leitmotif of my own career, and of this book, is that neuroscience—like other dynamic, disputatious fields ranging from cosmology to cancer research—often has as much to do with staking out territories, establishing priorities, and defending reputations as with furthering the cause of science. In these political maneuverings, scientists are generally no more self-aggrandizing than other ambitious people, but we are not necessarily less so, either.

Survival in a changing ecology requires more than a keenly competitive instinct, however. The forging of alliances between members of a species, or even between members of two vastly different species, is frequently the key to making a living in the natural world. Science, again, is no exception. These days it is rare indeed for a lone scientist to make an important breakthrough. Science has become too complex, and the growth of information too explosive, for truly independent research of the type that a Mendel or Darwin could once perform. What we see instead is interdependence—investigators, sometimes from very different fields, discovering a common ground of interest and a need to exchange expertise. The best of these collaborations results in a fresh perspective on an old, intractable problem, as happened in the case of opiate research.

The story I will tell in the following pages illustrates yet another kind of alliance—a type that, I suspect, has contributed more than any other to the advancement of science. The relationship I have in mind is the one between teacher and student in the laboratory. In the training of people who will conduct basic research, no classroom experience, textbook, or national conference can substitute for the daily opportunity to practice the tricks of the trade under the tutelage of a master scientist.

My own mentor at the National Institutes of Health, Julie Axelrod, was one of the best. His influence is apparent on practically every decision I have made in my professional life, including the way I approached the search for the opiate receptor. In relationships with students in my own laboratory, I have tried to emulate Axelrod. The opiate receptor story would have been very different if I had not.

· 1 ·

The Politics of Science

There are no more Medicis. Modern scientists are rarely if ever supported by wealthy personal benefactors. In the United States the vast majority of funds for biomedical research comes from the federal government through agencies such as the National Institutes of Health. Researchers are so dependent on this federal largesse that government policies often seem to determine the direction of scientific advances. When congressional appropriations for cancer research burgeon, the number of investigators for whom the mysteries of cancer hold fascination correspondingly swells.

But do more scientists spending more money on fashionable experiments automatically increase the chances of a genuine breakthrough? In fields such as high-energy physics, the answer is clearly yes; many crucial experiments cannot take place in the absence of multimillion-dollar equipment. And scientists in other fields, with comparatively modest needs, are usually more productive with twelve postdocs and technicians than with two. But once a certain threshold of personnel and funding is passed, one might argue that variations in resources are often only a minor determinant of what happens in a laboratory. With their grant money in place, most scientists seem eager to get on with the real work of adding their bit to the accretion of scientific knowledge, heedless of events in the less-than-pristine world outside the laboratory.

Some historians of science would thus discount the significance of government influence, arguing that the process of scientific discovery is inexorable, regardless of which way the political winds blow. If Watson and Crick had not discerned the lessons of the double helix, surely Linus Pauling, Rosalind Franklin, Maurice Wilkins, or some other worker in the field would have attained the same insight within a year or two. Researchers may maneuver their grant applications toward fundable directions, but their sights remain fixed upon a clear goal. According to this view, superficial investigations into what is momentarily fashionable never solve the "big problems."

Such questions and disputations about the wellsprings of scientific inventiveness fuel the daily thoughts of philosophers and historians. As a practicing researcher, I need no convincing. Politics matters in science. My own interest in opiates cannot be divorced from American political events in the early 1970s. In his first term as President, Richard Nixon confronted the apparent epidemic of heroin use in the United States, particularly among soldiers in Vietnam. Nixon's efforts to deal with the heroin problem played an inadvertent and indirect but nonetheless major role in the discovery of the opiate receptor and the subsequent revolution in molecular approaches to the brain.

Ever since the Harrison Narcotic Act of 1914 made heroin use illegal, this drug—a derivative of morphine—has been a chronic, endemic curse in the United States. Intravenous injection of heroin commenced among "bohemian" whites in the 1920s and after World War II began to infect the black community. Heroin was one of the ways young black males reacted to the bleakness of life in American big-city ghettos. For the forty years between 1920 and 1960, the level of heroin use seemed to have reached a plateau. This steady-state situation provided ample profits to the leaders of the Cosa Nostra and other underworld organizations that controlled narcotic traffic, while at the same time arousing only limited attention on the part of law-enforcement agencies.

Then something changed. No one is quite sure of the reasons, but there is ample documentation that the 1960s and early 1970s witnessed an extraordinary escalation of heroin use outside as well as inside ghetto cultures. As with many other illicit activities, it is difficult to obtain precise epidemiologic data—drug users are not exactly forthcoming when confronted by census-taking sociologists. One rough way to estimate the extent of heroin use is to count the number of people who die from overdoses. Some experts estimate that 1 percent of heroin users die each year from overdose. However, this "death" rule of thumb is not always reliable. During periods in which purer heroin preparations are available, undiluted with the usual large proportion of quinine or other adulterants, the death rate can increase precipitously for a period of time. I can attest to this phenomenon from a sad personal experience. In the early 1970s a bottle of pure crystalline morphine was stolen from my laboratory. Within three days, five patients were admitted to the emergency room at Johns Hopkins and other nearby hospitals with lethal or near-lethal overdoses. Police investigators were able to trace the source of injected drug to the theft from our lab. Of course, the drug culture quickly accommodates to changes in the purity of available heroin, and the death rate soon stabilizes.

For decades, New York City, especially Harlem, has had the highest density of heroin addicts in the country. In the late 1940s and early 1950s it was estimated that there might be 10,000–20,000 heroin addicts in New York. This calculation was based on a figure of about 100 deaths from overdose. Then, between just 1960 and 1964, the death rate more than doubled. It almost tripled between 1965 and 1970. The numbers increased still more rapidly in the early 1970s. In 1969 Dr. Michael Baden, the Associate Medical Examiner of New York City, estimated that New York City had roughly 100,000 heroin addicts.[1] Two years later most authorities were estimating that New York City possessed about 200,000 addicts. A rapidly rising crime rate was attributed to the growing number of heroin addicts seeking funds to support their habit.

Public concern about a heroin epidemic rose in the face of news reports that heroin use was spreading to younger and younger age groups. In 1960 only 15 teenagers were known to have died from heroin overdoses; in 1969 almost 250 died this way, and 20 of these victims were less than 15 years old. One of the 1969 heroin deaths was a 12-year-old boy named Walter Vandermeer, whose year or more on the streets supporting his habit was well documented by the *New York Times* in its depiction of "a heroin user who lived to be 30 in 12 years."[2]

But there was a greater concern to the voting public than the use of heroin among black teenagers and children in the ghetto. Heroin had begun to permeate the white middle class. The "flower children" of the hippie generation, who had started out with marijuana and LSD, were moving on to amphetamines and heroin.

Still more alarming to many people than the heroin epidemic at home was the extraordinarily widespread use of the drug by U.S. soldiers in Vietnam. By early 1971 the Defense Department estimated that between 10 and 15 percent of the 270,000 enlisted American soldiers in Vietnam were heroin users. Army officers with greater experience in the drug field felt that a more realistic estimate might be 25 percent, or roughly 60,000 enlisted men. In some combat units more than half the men were on heroin. Just as disconcerting as the numbers of soldiers involved was the nature of the heroin being consumed. In the United States, street heroin was only about 5 percent pure. In Vietnam, heroin was usually 95 percent pure. This, of course, greatly increased the likelihood of a user's becoming addicted.[3]

The epidemic of heroin addiction among soldiers bore particularly troublesome political implications for President Nixon. The American public became aware of the problem in early 1971, at the same time that they were becoming increasingly disgusted by the war itself. Atrocities, including the My Lai massacre, committed by American soldiers were being attributed at least in part to drug use by our troops. The spectre of hundreds of thousands of young American soldiers in Vietnam demoralized, drugged, and committing all sorts of monstrous acts was hardly the portrait of our fighting forces Nixon wished to convey when he argued for further bombings of North Vietnam.

Nixon's War on Heroin

On June 17, 1971, at the height of the undeclared war in Vietnam, Richard Nixon officially declared a war on drugs. He set up a special White House office to coordinate all Federal activities in the drug-abuse field, pulling together a wide range of efforts in disparate government agencies which among them were already expending about 500 million dollars a year. The Special Action Office on Drug Abuse Prevention, SAODAP, was to be closely directed from the office of the President. Nixon appointed as his "czar" of drug-abuse prevention Dr. Jerome Jaffe, a well-known psychiatrist with abundant experience in the treatment of heroin addicts. In a widely publicized press conference, Nixon, with his arms around Jaffe, announced his new offensive against heroin. He implicitly blamed the feuding of various Federal agencies for the failure to combat the heroin problem both at home and in Vietnam and exhorted Jaffe, as Director of SAODAP, to "bang heads together" and do whatever else may be necessary to "win this war." Never before in the history of the U.S. government had so much political power been accorded a physician.

At that time I had not conducted any research on opiates. I was not alone. Opiates were definitely not one of the glamour fields of brain research. To appreciate this situation, one must recognize the dilemma facing the research scientist. Unlike the President of the United States, an investigator cannot simply call in newspaper reporters and photographers, declare war on a scientific problem, then walk into the laboratory and solve it. To carry out a program of experimental research in a given area, the scientist must have, at a minimum, some strong hunches about what is going on and what experiments might be successful. In 1971 no one had any hard scientific information to explain exactly how opiates exert their effects on the human organism. Though a small cadre of narcotic researchers had been dealing with the problem of opiates for many years and had made some substantial contributions, they had not broken through to any fundamental understanding of the basic actions of the drugs. Accordingly, most brain researchers were at that time investing their major efforts in studies of other classes of drugs, such as antidepressants and amphetamines, where experimental strategies were clearer.

Though I hardly knew heroin from horseradish, I did know Jerry Jaffe. He and I were part of a rather small group of psychiatrists with an interest in biological approaches to mental illness. In the 1960s, when both of us trained, psychiatry was still dominated by psychoanalysis. A psychiatric trainee who expressed a strong interest in basic biological research was regarded as somewhat peculiar, perhaps suffering from emotional conflicts that made him or her avoid confronting "real feelings." An interest in science was regarded almost as sick, some sort of stratagem to avoid the psychoanalytic issues that *mattered* by fleeing to science. Jerry and I were both in this group of pariahs, but our approaches to the biology of mental illness were somewhat different. I chose to put most of my energies into basic laboratory research, whereas Jerry pursued clinical studies, becoming one of the pioneers in the use of the opiate methadone to treat heroin addiction. Based at the University of Chicago, Jerry transmuted the then experimental methadone maintenance approach into a large-scale treatment program involving a major portion of Chicago's heroin addicts.

When I saw Jerry on television and in the newspapers with President Nixon, I phoned to congratulate him and to offer whatever assistance I could provide. At this point, four days after the announcement of his appointment, Jerry had already moved into a small apartment in Washington and was preparing to fly to Vietnam the next day to review the drug problems of the armed forces. The President had promised the American public that he would make concrete inroads into the problem of drug abuse in the military within two or three months.

How was Jerry Jaffe, the fellow on the spot, supposed to accomplish all of these overwhelming assignments? True, he had the political clout of the White House behind any request he might make of other Federal agencies. But there remained quite a few roadblocks. To begin with, Congress had not yet passed legislation providing the administrative authorization for SAODAP and the funds necessary for this ambitious undertaking. Specifically, Jerry's offices occupied a lovely townhouse, sitting cater-corner from the White House, that had been elegantly restored by Jacqueline Kennedy's task force on Washington beautification. It made for a gorgeous home, but it was totally unfit for an office building. Jerry's dozen or so employees were cramped into the parlor, sitting room, and bedrooms of his lovely mansion-office.

Other problems abounded. Jerry's only mechanism for obtaining additional personnel was a presidential authorization that would enable him to transfer government employees from other agencies, especially the military, to the White House. He asked if I had any suggestions. At first I could think of nothing at all. I phoned him back when it occurred to me that a former medical student of mine, Alan Green, was at that time a commissioned officer in the Public Health Service doing research at the National Institute of Mental Health in Washington. I knew that Alan was dissatisfied: the laboratory environment was drab and the scientific problems under study bored him. More importantly, he always had a strong interest in the politics of medicine. I told Jerry about Alan. Within a week Alan was housed in the White House drug-abuse office as Jerry's special assistant. A few months later, with no more background than medical school and an internship, Alan was setting priorities for all Federally supported research in drug abuse in the United States, a one-hundred-million-dollar affair.

On occasions when I visited Alan and Jerry in their offices just a few doors from the White House itself, I was awed by the seeming closeness to power and the possibility of accomplishing almost whatever one would wish to accomplish. But I was equally impressed with the general chaos of such a hastily put-together hodge-podge of activities. In talking to Alan about his "research" responsibilities, I came to realize that much of it was a sham. Most of the so-called research really amounted to support of treatment clinics. Little of the funds went to scientific investigations of how opiates actually work in the body. Treatment clinics at that time focused mostly on administering methadone to addicts. This hardly represented a definitive solution, as it merely substituted one addictive opiate for another. Of course, methadone is given by mouth, not by injection, and is longer-acting than heroin. More importantly, it is legal and so goes a long way toward lessening the multibillion-dollar cost of crime associated with street heroin every year. Still, hundreds of times more money was being pumped into the drug clinics than into basic research. One could double the research expenditures without hampering the clinical work more than one or two percent. Thus, why not support basic research aimed toward a fundamental understanding and ultimately a true resolution of the addiction problem?

The political reasoning behind a policy to put money into clinics and not science was elementary. The White House was already planning Nixon's 1972 reelection campaign. Heroin was a politically hot area. If the President could announce major reductions in addiction achieved in only a few months, he could defuse a potential target for Democratic assault. Solutions rarely emerge so rapidly or predictably from scientific attempts to unravel basic biological processes.

While political considerations may have been paramount in the eyes of White House aides, Nixon himself seems to have possessed a broader perspective. Jerry has related to me that in the course of a meeting at San Clemente, the President personally told him "to make certain that something of long-range value emerged from the effort." History plays ironic tricks. According to Jaffe, other events with long-range ramifications took place in the same house within a few hours of his discussion with the President. It was July 2, 1971, the same day Nixon told Egil Krogh to set up the "plumbers." (See figure 1.)

Letter to the White House

Basic scientific research is chronically underfunded by the federal government, whether it involves opiates, allergy, or

Figure 1. Dr. Jerome Jaffe, Special Consultant to the President for Narcotics and Dangerous Drugs, reports to President Nixon on the lawn of San Clemente. (Photo courtesy of UPI/Bettmann Newsphotos.)

even cancer. One day in early October I was bemoaning this state of affairs to a friend, Arnold Mandell, then chairman of the Department of Psychiatry at the University of California at San Diego. I told him about the massive amounts of money emanating from the White House in the drug-abuse area, with so little going to scientific activities. It seemed to me that Jerry Jaffe could, with relatively little cost and effort, set up a group of drug-abuse research centers, where the best minds in biomedical research could investigate the mechanisms in the brain that underlie drug addiction. I had even suggested this strategy to Jerry himself in conversations during the summer, but he had seemed preoccupied by the more pressing battles with the Defense Department as to who was the real boss of treatment programs for drug-using soldiers—the Armed Services or Jerry Jaffe. I had also described the proposed center program to Julius Axelrod at the National Institutes of Health. He was sympathetic but had no ideas as to how one might influence government to implement the proposal.

I was fairly disheartened and felt that moving money into basic research centers was a hopeless venture. Indeed, I am generally pessimistic when I have to deal with massive governmental or corporate bureaucracies. The experience leaves me feeling a bit like Sisyphus in Hades, endlessly rolling a weighty rock up the hill, only to watch it roll back down again just as I reach the summit. Arnie, on the other hand, was an irrepressible enthusiast, frenetically full of novel ideas, some of which were unrealistic fantasies, while occasional proposals, just as original, were eminently feasible.

We bandied back and forth a variety of possibilities. Finally, Arnie suggested, "Why don't we formally request that Jerry set up and fund those research centers and that he mandate that they focus on the basic questions?" I reminded Arnie that Jerry had many other things to worry about and was already working eighteen hours a day. Our importuning him would only add yet another unwelcome stress. Certainly he would view our imposition as taking advantage of friendship. Arnie retorted, "But Jerry must realize surely that these daily political wars won't accomplish much of anything in the long run. If he wants to be remembered, he'd better set in motion some program that will last longer than his personal government service."

I was still dubious. However, Arnie and I collaborated in drafting a letter to Jaffe proposing the creation of a limited number of drug-abuse research centers dealing with both basic scientific and clinical questions about the nature of drug abuse. We were not altogether selfless in our proposal. We suggested that Hopkins and UC San Diego might be excellent locations, among others, for such research centers. I worried that a letter to Jerry might never be seen by him, buried under the avalanche of daily mail. Accordingly, I hand-delivered a copy to Alan Green, who spent three or four hours each day at Jerry's side. At Arnie's behest, we addressed the letter jointly to Bert Brown, Director of the National Institute of

Mental Health, and to William Bunney, then Director of the Division of Narcotic Addiction and Drug Abuse within the National Institute of Mental Health. The letter was dated November 12, 1971.

Addressing the letter to Bunney as well as Jaffe turned out to be a crucial step. Bunney was and still is one of the country's leaders in clinical studies of biological psychiatry. His areas of special expertise are depression and mania. Though Bunney had negligible background in drug-abuse work, he was selected to head NIMH's drug-abuse unit, apparently to lend his international prestige to the effort. As Bunney was a research psychiatrist, we hoped he would be favorably disposed toward our proposal. However, since Bunney was not himself involved in opiate studies, we were not sure that he would expend his few political chits on our enterprise.

Bunney surprised us. He knew well that the Presidential reelection strategy in the drug-abuse area called for treatment clinics but not science. He knew far better than Arnie or I the inner workings of White House agencies. We thought that our sole hurdle was Jerry Jaffe. Naively, we assumed that Jaffe was indeed the czar of drug abuse, with truly autocratic control over the federal drug-abuse effort. In fact, however, his powers were extremely limited. The Nixon White House kept firm and close control over all government activities that "mattered." And in 1971 the heroin problem was a particularly visible, high-priority item. Consequently, SAODAP fell under the mantle of John Ehrlichman's Domestic Council. Because of heroin's political ramifications, Bud Krogh, architect of the plumbers unit and one of Ehrlichman's most trusted aides, personally supervised all SAODAP activities and possessed an absolute veto power over major policy initiatives.

I suppose it is not altogether surprising that Jaffe's activities were so closely monitored. The detailed oversight, however, extended also to Bunney. Almost all the research dollars in drug abuse flowed through his division. Bunney's interest in research, and his regular association with scientists who were almost all liberal Democrats, had already engendered White House suspicions as to his loyalties. Arnie and I knew there existed some shadowy opposition to spending drug-abuse monies for research, but we never dreamt how complicated matters really were in the White House labyrinth. We assumed the bottleneck was someplace at SAODAP, and we hoped Bunney could help out. As we were to learn much later, advocating our case did place Bunney at some serious political jeopardy. However, he accepted whatever risks might be involved and strongly supported the center concept.

Unbeknown to Arnie and me, support for a center type of approach was already developing in the government even before we sent our letter. Hoping to pry loose some of the massive drug-abuse dollars for basic science, Bunney in September had brought together a group of biomedical researchers working at NIH to brainstorm ways of efficiently using Federal monies to solve the important underlying scientific questions in drug abuse. Julius Axelrod was among the group. He described the center notion that we had discussed back in the summer, emphasizing the importance of using the centers as a means to attract new people from other areas into drug-abuse work. Ideally, he argued, one should go out and deliberately recruit the best biomedical scientists, regardless of their previous research interests. Axelrod's idea was essentially to do for the drug-abuse field what focus on the atomic bomb had accomplished for nuclear energy—to accelerate scientific evolution tenfold so that discoveries which normally percolate over several decades might erupt in two or three years. One would not likely attain such a goal with the same small group of opiate researchers who had been progressing, but fairly slowly, since the turn of the century. An infusion of new perspective from workers in other areas was crucial.

Axelrod's ideas prevailed, and Bunney commenced to lobby for research centers as the cornerstone of Federal support for basic scientific investigations into the mechanisms of drug addiction.

Drug-Abuse Research Centers Announced

I still do not know details of the ensuing events, but whatever transpired moved briskly. On December 23, 1971, just over a month after the letter from Arnie and me describing the center proposal, another letter emerged under Bunney's signature, directed to scientists in a wide range of areas, including many who had never before worked on drug-abuse questions. It invited them to compete for one of four anticipated drug-abuse research centers. Two million dollars were allocated for the center program, an extraordinarily modest fraction of the total Federal budget allocated to drug abuse. The *angst* required to free up this pittance reflects how tightly Nixon's people controlled monies.

In any event, that the research program existed at all was promising, even though there was only enough money for a handful of centers. The rapid pace of developments was accentuated by the proposed timetable. In most Federal agencies, money not expended before the end of the fiscal year is essentially lost. For center applications to clear the complex external and in-house governmental review process before July 1, 1972, required unprecedented expedition. Bunney's letter of December 23 demanded that applications be received no later than February 1, 1972. Considering the time lost during Christmas vacation as well as delays caused by university red tape, it seemed that we would have to conceive, compose, and complete a several-hundred-page grant application involving dozens of scientists in about two weeks.

We did it at Hopkins. Almost two dozen other universities were also able somehow to meet the deadline. The government then quickly assembled an ad hoc review committee of scientists knowledgeable in areas related to drug abuse. The review group of scientists emptied their schedules of other commitments and traveled about the country evaluating each of the applicant organizations. By late spring the decision-making process was complete. Rather than establishing only four centers at $500,000 annually for each, the reviewers chose instead to spread the funds more thinly, establishing six research centers. Johns Hopkins and the University of California at San Diego were included.

Is scientific discovery always predicated on political events such as this grant-getting competition? There is no simple answer. But one important way in which politics impinges on working researchers, at least upon me, has as much to do with emotions as with the actual funds that I receive. If a committee of my scientific peers at the National Institutes of Health awards me a large grant, I experience a rush of self-esteem—the reviewers might just as well have administered an infusion of cocaine. I commence to think more highly of my abilities, to value my own ideas, and therefore to pursue them more aggressively. Maybe the ideal scientist would have such a deep inner assurance that he or she could push forward stubbornly, regardless of the resistance or indifference of peers. But I am far from this ideal; I suspect many other scientists also fall short.

Political forces can direct one's thinking processes as well as one's emotions. Of course, the most original conceptions often seem to well up into consciousness from nowhere at all. Many scientists have undergone these exhilarating, almost unnerving, experiences. Yet such insights invariably represent syntheses of material mulled over consciously again and again in the preceding weeks and months.

Before 1971 I had devoted negligible attention to opiates and consequently had formulated few thoughts, and fewer insights, into their actions. But Jerry Jaffe's almost messianic exhortations to "give me a hand" and his repetition of the immense challenges posed by the heroin epidemic and the comparably great opportunities available if only the best minds would help out—these words were not without an intellectual as well as an emotional effect. I waded through the chapter on narcotics in Goodman and Gilman's standard pharmacology textbook, a chapter authored by Jaffe himself. What immediately struck me was the absence of any genuine understanding of how opiates act. Why does morphine relieve pain? Why do opiates cause euphoria? Why do all opiates depress breathing? Respiratory depression is the most serious side effect of these drugs, accounting for most deaths from overdose. I spent weeks in the Hopkins library surveying the past hundred years of opiate research, but I gained only the most fragmentary insight into these questions. Even the relatively minor, well-circumscribed effects of opiates were largely unexplained; for example, why do opiates constrict the pupils of the eyes? This is such a uniform response to the drugs that police officers regard pupillary constriction almost as proof positive when confronted with a potential addict, especially one wearing a long-sleeved shirt in the middle of summer, to hide the track marks left by needles. No one could explain adequately why opiates cause nausea or why they are so effective as cough suppressants. And, of course, the textbooks and published papers offered nothing but untested and often untestable theories as to the causes of opiate addiction or, for that matter, addiction to any substance.

A perusal of the types of experimental studies published suggested why there had been so little progress. Simply put, no one was attacking fundamental issues. The situation was analogous to what confronted James Watson when he wished to understand how genes regulate heredity. Watson was well trained in genetics and thus fully aware of the fundamental dilemmas in the field. Possessed of a critical intellect and far-reaching vision, he realized that most of the geneticists were simply not asking the right questions. Watson appreciated that one could not hope to explain in a universal fashion genetic mechanisms without working out the structure of the genetic material, DNA: "That was not to say that the geneticists themselves provided any intellectual help. You would have thought that with all their talk about genes they should worry about *what they were*, yet almost none of them seemed to take seriously the evidence that genes were made of DNA. This fact was unnecessarily chemical. All that most of them wanted out of life was to set their students onto uninterpretable details of chromosomal behavior."[4]

The situation was similar in the opiate field. Pharmacologists had failed to ask explicitly the question, "Where is the site of opiate action?" They had characterized in great detail all manner of effects wrought by opiates in animals. At a biochemical level, they would administer morphine and measure its effects upon proteins, nucleic acids, carbohydrates, lipids, and every neurotransmitter imaginable. Opiates did influence almost whatever parameter was measured. But the really important task is to seek the locus of the *initial* action of the drug, that site affected first and universally by all opiates, the site whose activation then triggers secondary changes in diverse physiological systems.

By simple analogy, assume that someone has been inflicted with a fatal hammer blow to his head. However, the coroner is blind and so cannot simply observe the victim's body. Instead he must employ chemical tests to explain the death. Being a careful chemist, he will seek and find alterations in everything he measures, down to the victim's toenails. But all that chemical data may not answer the basic question, what killed him? The coroner would have to be fairly clever to deduce from the indirect chemical information that the initial insult was a blow to the head. I had a vague feeling that many opiate pharmacologists, like the classic geneticists Watson satirized, had been groping in the wrong directions. There should be a simpler strategy to find the initial site of opiate action.

Back in the summer of 1971 I had attended a Gordon Conference on molecular pharmacology. Gordon Conferences, originally established in the 1920s by Dr. Neil Gordon of Johns Hopkins, are relaxed scientific meetings usually held at a variety of prep schools or junior colleges scattered through New England. An air of informality is maintained by restricting the number of participants to a maximum of 150, proscribing any publication of the proceedings, and scheduling far more time for open discussion than for formal presentation of papers.

By 1971 the notion that answers to how drugs act might best be sought at a molecular level was gaining currency, even among classical pharmacologists. Accordingly, the meeting on molecular pharmacology was both well attended and spirited. Most of the presentations dealt with topics that were readily amenable to molecular approaches. For instance, there were numerous papers showing how antitumor drugs block multiplication of cancer cells by insinuating themselves into the layers of the DNA double helix, thus preventing genes from duplicating themselves.

One talk especially attracted me. Avram Goldstein, Professor of Pharmacology at Stanford, spoke about his attempts to measure the initial site of action of opiate drugs, the putative opiate receptor. Technical difficulties were preventing him from identifying the specific sites, but it was apparent to me that Goldstein was approaching the problem in the most sensible way I had yet perceived in the narcotic field. I took more notes on his lecture than on the other two dozen talks combined.

Grantsmanship

Preparing the application for the drug-abuse-center grant had consumed all of Christmas vacation and the succeeding four weeks. Successful grant-writing is an art form all to itself. One of several secrets is not to describe a complete, original, as yet untested idea, even one which you think may lead to a tremendous breakthrough. Such an application is likely to be rejected straight out. The reasoning of the grant referees goes something like this: Any novel, untested idea is by definition a long shot. It is at least as likely to fail as to succeed. If the government has awarded you half a million dollars annually for five years and the pilot experiments blow up, the granting agency will have lost a most costly wager.

A far more successful strategy for obtaining grant support is to propose to do something you have already done. Thus, if you have already discovered some phenomenon of nature, you would do well to describe the finding in your application and then propose to devote the succeeding five years to follow-up research, characterizing the phenomenon in detail, replicating experiments, extending the results to other animal species, and so forth. Such an approach may seem a trifle tedious, but it represents the bread-and-butter, yeoman service of science. Most importantly, if the initial finding that you propose to extend is of reasonable importance, then the granting agency can feel confident that its money will not be squandered in dead-end efforts.

From 1968 through 1971 I had been studying the neurotransmitters norepinephrine and dopamine (see Chapter 9). These chemical messengers of the brain play prominent roles in regulating emotions. Abundant evidence from numerous laboratories, including our own, indicated that enhanced action of norepinephrine and dopamine was responsible for the euphoria and alertness that comes from use of amphetamines. Amphetamines, of course, were being widely abused in the late 1960s. In the streets of the Haight-Ashbury district of San Francisco, thousands of amphetamine addicts daily injected themselves intravenously with preparations variously labeled speed, crystal, or dex. Intravenous users—speed freaks—often built themselves up to daily doses hundreds of times the amount they first required to get high. Investigation of how amphetamines act certainly qualified as bona fide, respectable, and predictably successful drug-abuse research. We had already published several papers on our initial findings, so that an application proposing to carry the work forward along relatively obvious lines would not likely be attacked. There was only one minor difficulty: By late 1971 I was becoming quite bored with amphetamines.

What I really wanted to do was search for the opiate receptor. I had a few ideas as to how I might overcome the technical problems that had haunted Avram Goldstein's forays in the area. But I had no pilot results auguring success. I had not performed even the most preliminary experiments. Any experimental work would require the custom preparation of radioactive chemicals, a process that might consume several months, and we had only a few weeks to meet the grant application deadline.

I compromised. In the application, most of the proposed experiments represented a straightforward exploitation of leads we had already obtained as to molecular sites of amphetamine action. A final, relatively short section of the application described the proposed strategy to identify opiate receptors.

When the scientific referees came to Hopkins for a site visit, the scenario played out so predictably that it seemed to have been scripted in advance. The reviewers applauded the amphetamine proposal as "real solid stuff" and funded it fully. The opiate-receptor project was described as "a most risky flyer," whose proposed funding was reduced considerably.

I knew, of course, that NIH regulations permit a researcher considerable flexibility in how he handles his grant money. Once the grant was awarded, with a starting date of June 1, 1972, I dropped the amphetamine work abruptly and commenced attempts to find the opiate receptor.

· 2 ·

Drug Wars

I never do laboratory experiments myself. Being a colossal klutz, I genuinely worry that I will drop something and cause an explosion that will propel broken glass in a hundred different directions and kill half the people in the room. All hands-on experiments in my lab are carried out by students. I work closely with each of them, scrutinizing their experimental results and brainstorming at length about what to do next.

Of all these activities, the weightiest decision for me is the initial selection of an experimental direction for each student. This decision emerges in part from my pile of "ideas for students." Running a laboratory with eight to ten PhD candidates and postdocs requires a steady storehouse of ideas for interesting new projects. Roughly nine out of ten experimental efforts that we undertake fall flat on their face. To keep ten students in business at any given time demands a backlog of almost 100 new ideas.

Pulling the appropriate project from the pile of paper scraps in my "idea" basket depends in large part on my judgment of the individual student. Some students are extraordinarily systematic and fastidious. They do best with projects that demand close attention to detail. These young men and women tend not to worry too much about the inherent glamour of their effort. They also are less concerned with whether they are working on a completely novel or slightly worn project. By contrast, there are other bright students who pay less attention to detail, who are more inclined to think in terms of the "big picture." They may be reluctant to work in an area unless they can foresee its broadest implications. When I choose a research direction to suggest to a student, I try to give some thought to the fit between the project and the personality.

Candace Dorinda Bebe Pert entered graduate school at Hopkins in the Pharmacology Department in 1970. She came in large part because her husband, Agu, was a PhD psychologist doing his military service at the Chemical Warfare Center at Edgewood Arsenal, a suburb of Baltimore. Candace had a somewhat rococo background. A native of New York, she had attended Hofstra, a small college on Long Island where her family lived. While at Hofstra she met Agu. Candace quit school to marry Agu and move to Philadelphia, where he was to pursue his PhD in psychology at Bryn Mawr College. There she soon gave birth to their first son, Evan. With no family wealth to back them, Candace and Agu were forced to survive on little. There was certainly no possibility of Candace's returning to school. Indeed, besides caring for her small baby, Candace had to seek work. Determined to make the most money possible in the time she had set aside for work, Candace took a job as a cocktail waitress. And being naturally garrulous, Candace often engaged her customers in lighthearted but sometimes serious banter. One restaurant guest turned out to be an assistant dean at Bryn Mawr College. Perceiving that Candace was far from a typical cocktail waitress, she encouraged her to complete her education, and even personally arranged for interviews and then admission to Bryn Mawr. Shortly after her graduation, Candace and Agu moved to Baltimore.

Initially, Candace had some difficulty in adjusting to the more rigorous academic demands imposed upon graduate students. At one point the department's senior faculty admonished me to structure her environment so that she might proceed more systematically through the required courses and laboratory work. Following the orders of my colleagues, I set up a program of research for Candace on a well-delineated project that could be approached systematically with great likelihood of success. She was to explore mechanisms that regulate the production of acetylcholine, the longest known and most prominent neurotransmitter in the body. Neurotransmitters are chemical messengers released by neurons (nerve cells) to excite or inhibit adjacent neurons in the brain or, outside the brain, to regulate the heart, intestine, and numerous other organs. Acetylcholine had been discovered in the early twentieth century as the neurotransmitter that slows the beating of the heart. In the brain, it plays an important role in memory circuits; and when a specific group of acetylcholine neurons begin to degenerate, the result is Alzheimer's disease.

The experimental approach I proposed to Candace augured success, because it was basically a follow-up to some exciting discoveries made by Dr. Henry Yamamura, a postdoctoral fellow who had been working with me. Hank had found that the formation of acetylcholine was not regulated by the enzyme that converts its precursor, choline, into acetylcholine. Instead, Hank discovered that what mattered most was a unique pump that sucked the precursor choline into the acetylcholine-containing neuron. Hank had identified this pumping mechanism in the brain. If the choline pump did indeed regulate the function of acetylcholine-containing neurons, then we should try to link this pump directly to some function of the neurons.

Unfortunately, measuring a functional consequence of activity in a selected group of neurons was virtually impossible in the brain. The brain is so crowded with literally billions of neurons that one cannot focus readily on a single population for functional studies. The intestine, which is also rich in acetylcholine neurons, is another matter. Here, they regulate intestinal contractions that we could easily measure in the laboratory, using small strips of intestinal muscle with attached nerves. I wanted Candace to characterize the choline pump in intestinal strips so that we could relate the pumping of choline, the subsequent formation of the neurotransmitter acetylcholine, and the actual contracting function of the intestine.

This seemed to me to be a fascinating research program, with a successful doctoral dissertation virtually guaranteed. Many graduate students would have been thrilled at the possibility. Candace was not. She felt that I was inflicting her with Hank Yamamura's hand-me-downs. Though she complied grudgingly, Candace set no world records for enthusiasm or experimental progress.

In the summer of 1972, at a point where she had accomplished a certain modicum of respectable work, I knew it was time to terminate this rather dull liaison of student and project. I suggested that we take what research she had completed on acetylcholine, assemble it for publication in a modest journal, and go on to something new for her doctoral work. I proposed to her the idea of the opiate receptor.

Candace was ecstatic—a project of her very own. We set to work immediately.

Opium among the Ancients

For the next few months Candace and I would immerse ourselves in the search for the opiate receptor. Laboratory days of 12 hours or more would become the rule rather than the exception. In addition, I had fallen into a habit of spending many of my evenings and weekends trying to learn what was already known about opiates—not just about their molecular properties but about the history of opiate use and addiction. I was becoming fairly obsessed with a subject about which, just a year before, I had known practically nothing.

In my reading I learned that written references to plant extracts that produce effects sounding very much like opium have been unearthed in what was then called Sumeria—today, southern Iraq—as far back as 4000 B.C. But how did early humans first discover that extracts of the poppy plant affect the mind, I wondered. Such a question is not just of academic historical interest. My curiosity had an ulterior motive. I knew that, despite all of our powerful new chemical tools, most of the major drugs used in medicine today stem from the herbal extracts of folk medicine. Thus, digitalis, which relieves congestive heart failure and is one of medicine's most life-saving drugs, was first an extract of the foxglove plant, *Digitalis purpurea*. Similarly, the belladonna plant, *Atropa belladonna*, gave rise to the alkaloid atropine, which blocks effects of the neurotransmitter acetylcholine and has many uses in medicine. The medical children and grandchildren of atropine have provided all our presently employed antihistamines, antischizophrenic, and antidepressant drugs. From the coffee bean comes not only caffeine but also theophylline, the most widely used agent in treating asthma. Fungal plants have given rise to ergot drugs, of importance in treating migraine headaches, contracting the uterus to initiate labor, and many other uses. And of course the first antibiotics, from penicillin through tetracycline and streptomycin, all were extracted from common fungi. Perhaps the ways in which early civilizations harvested poppy extracts and used them in herbal medicines held clues about how more effective, less addictive opiates might be developed.

That someone should have discovered the psychoactive properties of the opium poppy, *Papaver somniferum*, is remarkable if we consider that it is only one of hundreds of species of poppy plants, most of which have no effect on the mind. The ancient medicine woman or man who first identified the psychic actions of opium also had to deal with the subtle timing required to obtain morphine—the active ingredient in opium—from the poppy. The poppy is an annual plant. Morphine is generated only between the time the petals of the plant drop and the seed pod matures. On just the right day the plant must be harvested, and in the early evening a shallow cut must be made into the interior of the relatively unripe seed pod, so that during the night the raw opium—a white, milky substance that contains morphine—will ooze from the slits. By morning it turns reddish brown and sticky, a little like caramel; this opium resin is immediately scraped from the pods and compressed into small, firm blocks or balls, which can be smoked or swallowed.

The best documentation of the early use of opium stems from classical Greek culture. In the *Odyssey*, written about 1000 B.C., Homer describes a plant-derived drug, nepenthe, whose properties very closely resemble those of opium. The fact that the poppy plant produces blissful somnolence and sleep was apparent from numerous references in Greek and Roman literature, particularly in the writings of Ovid and Virgil. Our medical term for a sleeping pill, *hypnotic*, comes from the Greek god of sleep, Hypnos. In most depictions of Hypnos in Greek art, he is shown carrying with him bundles of poppy plants. The Roman god of sleep, Somnus, from whose name we derive the term *somnolence*, is often shown carrying a container filled with the juice of the poppy plant.

Early Greeks were well aware of the recreational potential of opium, but they appreciated its medical virtues as well. The writings attributed to Hippocrates describe a plant that presumably represented opium. Clear-cut written documentation explaining in detail how to prepare opiate extracts and employ them in specific medical circumstances first appeared a few decades after the death of Christ. The rather convoluted procedure has been well known and employed commercially for almost 2000 years. Scribonius Largus wrote one of the earliest systematic pharmacopoeias, *Compositions Medicamentorum*, in the year 40 A.D., in which he described in some detail how one incises the poppy capsules to obtain the therapeutically efficacious juice. Dioscorides in 77 A.D. conducted fairly well-controlled clinical pharmacologic studies comparing effects of whole plant extracts versus opium juice and established that the active ingredient occurs only in the juice of opium poppies at a particular time of development.

The Greek word *opius* was coined about this time. It literally means "little juice" and presumably refers to the small amount of precious raw opium that can be extracted from a single poppy plant. In a fashion not too dissimilar from present-day medical fads, Greek physicians employed opium almost as a panacea. The eminent Greek physician Galen wrote that opium "cures headache, dizziness, deafness, epilepsy, poor vision, asthma . . . tightness of breath [perhaps congestive heart failure], colic, jaundice, stone [presumably gallbladder], urinary disorder, fevers, dropsy, leprosy, the troubles to which women are subject, and melancholy." Interestingly, some of these uses fit well with what would today be considered valid indications for prescribing opiates. Galen clearly appreciated the ability of opium to relieve the pain of headache, gallbladder colic, and kidney stones. Relief of cough is still an important use of opium-derived drugs. Most people are not aware that almost all cough medicines, even over-the-counter remedies, rely on the opiate-related chemical dextromethorphan for their effectiveness. Some of the other uses Galen described probably reflect the calming effect of opium, which by itself would ease the agitation that worsens asthma and gynecologic disorders.

With the fall of the Roman empire, advances in medicine came almost exclusively from Arab physicians. Because Islam prohibits the use of alcohol, opium became widely used as a "safer" recreational agent. It also had an important impact on Arabic medicine. The writings of Arabian physicians formed the basis for the subsequent development of modern European medicine. Avicenna, the most celebrated of them, considerably improved the ways different types of opiate extracts were administered to patients with specific disease entities. He worked out a method for treating diarrhea with opiate extracts that is still in use today. Avicenna himself died of a presumably accidental overdose. His death highlights one of the severe limitations of medical practice prior to the era of scientific chemistry. In those days, physicians never had available a pure chemical entity of known potency. The amount of active substance might vary five- to tenfold in different opiate extracts. Avicenna was apparently the victim of a much-purer-than-usual batch. Such miscalculations account for most street heroin deaths today and provide a strong argument in favor of making heroin legally available to addicts.

Product Development in the Renaissance

In the sixteenth century, European medicine began to incorporate the lessons of the Greeks, Romans, and Arabs. Paracelsus, the most influential of the Renaissance physicians, was an advocate of opium as the most universal of medical treatments. He was quoted as saying that opium "will dissolve disease as fire does snow."

Paracelsus owed his therapeutic success with opium to his vigorous administration of high doses. Increasing the dose of a drug might seem fairly straightforward. However, in those days, before the advent of experimental scientific medicine, altering the dose of a drug could be a complex matter. Not only did opium batches vary greatly in their content of morphine, but opinion as to the identity of the active ingredient in therapeutic preparations was muddled at best. Opium had to be extracted by a solvent, frequently containing alcohol, to be clinically effective. Over the years physicians gradually added one, then another, ingredient, often a spice, to improve the therapeutic effect. With strong spices, patients could sense directly that they were receiving something "active." As the list of additives slowly grew, physicians sometimes lost track of just which component was securing the therapeutic effect.

One of the most widely used medications employed literally as a panacea for all sorts of diseases was called theriaca. It was developed by Andromachus, physician to the Roman emperor Nero, and was used at least until the late eighteenth century. Galen's recipe for theriaca was as follows:

> Root of Florentine iris, licorice, 12 ounces each; of Arabian costus, Pontic rhubarb, cinquefoil, 6 ounces each; of Ligusticum meum, rhubarb, gentian, 4 ounces each; of birthwort, 2 ounces; herb of scordium, 12 ounces; of lemon grass, horehound, dittany of Crete, calamint, 6 ounces each; of pennyroyal, ground pine, germander, 4 ounces each; leaves of laurus cassia, 4 ounces; flowers of red rose, 12 ounces; of lavender, 6 ounces; of St. John's wort, 4 ounces; of lesser centaury, 2 ounces; saffron, 2 ounces; fruit of amyris opobalsamum, 4 ounces; cinnamon, 12

> ounces; cassia lignea and spikenard, 6 ounces each; Celtic nard, 4 ounces; long pepper, 24 ounces; black pepper and ginger, 6 ounces each; cardamoms, 4 ounces; rape seeds, agaric, 12 ounces each; seeds of Macedonian parsley, 6 ounces; of anise, fennel, cress, seseli, thlaspi, amomum, sandwort, 4 ounces each; of carrot, 2 ounces; opium, 24 ounces; opobalsamum, 12 ounces; myrrh, olibanum, turpentine, 6 ounces each; storax, gum arabic, sagapenum, 4 ounces each; asphaltum, opoponax, galbanum, 2 ounces each; juice of acacia and of hypocist, 4 ounces each; castor, 2 ounces; Lemnian bole, calcined vitriol, 4 ounces each; trochiscs of squill, 48 ounces; of vipers, of sweet flay, 24 ounces each.
>
> Triturate the balsams, resins and gums in a sufficient quantity of wine to form a thin paste, and incorporate the whole with 960 ounces of honey.[1]

Three other standbys in use as long as theriaca were mithridatium, philonium, and diascordium. These four "officinal capitals" were presumed to be distinct entities with different therapeutic indications. A careful examination of their contents, however, shows that the one item all four held in common, and in substantial amounts, was opium.

Paracelsus had the insight to appreciate, either explicitly or implicitly, that opium itself was the key ingredient. He developed a new preparation, laudanum, that was a far simpler, more potent, and direct alcoholic extract of the opium poppy. Then in the late seventeenth century, Dr. Thomas Sydenham, one of England's most eminent medical practitioners and writers, developed a more effective form of laudanum. His recipe contained a dram of cinnamon, an ounce of saffron, and two ounces of opium mixed together in a pint of Canary wine. On the therapeutic effects of opiate drugs Sydenham wrote: "I cannot forebear mentioning with gratitude the goodness of the Supreme Being, who has supplied afflicted mankind with opiates for their relief; no other remedy being equally powerful to overcome a great number of diseases, or to eradicate them effectually."[2]

Several similarly crude opiate extracts are still widely used today. One of the preferred treatments for diarrhea is paregoric, which contains, besides opium and alcohol, only anise oil, benzoic acid, and camphor. It is extensively prescribed in the United States and all over Europe. (I take a bottle with me on all trips to France.) But crude plant extracts make for notoriously unreliable medicines. The content of the active ingredient in a plant varies enormously from strain to strain and even between different plants of the same strain grown during different types of weather. The preparation process also affects the amount of active ingredient that emerges. Accordingly, one of the major advances in modern medicine has been the isolation of active chemicals from a variety of plants. The first important isolation of an active alkaloid (a nitrogen-containing drug) was achieved with the identification of morphine in the poppy plant.

Credit goes to a twenty-year-old German chemist named Frederich Sertürner. His first experiments, published in 1805, showed that fully 10 percent of the poppy plant by weight consists of morphine. For the following ten years he continued to characterize the chemical and pharmacologic properties of pure morphine, using himself as a guinea pig in most instances and coming close to killing himself from overdose in the process. Sertürner's definitive account of the principal active chemical in opium, which he named after Morpheus, the god of dreams, was widely acclaimed. The Institute of France awarded him its most prestigious prize, along with 2,000 francs, a small fortune in those days. Isolation of other active ingredients from the poppy followed in short order. In 1832 the eminent French chemist Robiquet isolated codeine. In the same year another Frenchman, Pelletier, obtained from the poppy plant the opiate thebaine, which is the chemical starting point for synthesizing large numbers of opiate derivatives used in medicines today.

Besides its importance for the opiate field, Sertürner's isolation of morphine was one of the major breakthroughs in drug chemistry. A host of revolutionary new drugs quickly followed. Pelletier, the chemist who had isolated thebaine, extracted quinine, so important for treating malaria, from tree bark. Pelletier also discovered the convulsant drug strychnine, used now mostly in rat poison but also of importance as a research tool for clarifying neuronal activity in the brain. In the mid-nineteenth century cocaine was isolated from coca leaves. As the first local anesthetic, cocaine vastly expanded the therapeutic potential of surgical operations and gave birth to the entire field of eye surgery. Digitalis from foxglove and atropine from belladonna were also isolated during this time.

From Use to Abuse

With such widespread acceptance and prescription by the medical profession, opiates were splendid candidates for abuse. That the European medical profession knew well the addictive properties of opiates in the sixteenth and seventeenth centuries is evident in the description by Dr. John Jones of the nature of the addictive process: "The effects of suddenly leaving off the uses of opium after a long and lavish use thereof are great and even intolerable distresses, anxieties and depressions of spirit, which commonly end in a most miserable death, attended with strange agonies, unless men return to the use of opium; which soon raises them again and certainly restores them."[3] However, major social problems with opiate addiction did not emerge in Europe until the nineteenth century.

The British writer Thomas DeQuincey first used opium, in the form of Sydenham's laudanum extract, to ease the pain of a toothache. "That my pains had vanished was now a trifle in my eyes . . . Here was the secret of happiness, that which philosophers had disputed for so many ages, at once discovered; happiness might now be bought for a penny, and carried in the waist-coat pocket; portable ecstasies might be had corked up in a pint bottle; and peace of mind could be sent down by the mail."

DeQuincey became a thoroughgoing laudanum addict. He introduced several other British writers to the drug, including Samuel Taylor Coleridge and Elizabeth Barrett Browning, who themselves became addicts. Coleridge's famous poem *Kubla*

Khan was the description of a vision he had while under the influence of opium. "The Confessions of an English Opium Eater," DeQuincey's hymn in praise of opium, published in 1821, inaugurated a half-century of proselytizing by many prominent figures in the British and French literary community. The transcendence made possible by opium was popularized in much the same way that tripping on LSD was glorified by Timothy Leary in the 1960s.

An important event in medicine, and in the use and abuse of opiates, was the development of the hypodermic syringe in 1853 by Dr. Alexander Wood. Before the advent of the hypodermic syringe, even pure drugs such as morphine had to be taken by mouth. Injectable drug treatment in medicine commenced with the use of the syringe. Analgesia following injected morphine could be obtained far more reproducibly and rapidly than with orally administered opium. These advantages became particularly crucial during the American Civil War, the Franco-Prussian War, and the Prusso-Austrian War.

Many Civil War veterans returned home as addicts to injectible morphine, and this "soldier's disease" was blamed by some for the high level of opiate addiction that was rapidly developing in late nineteenth-century America. But no such epidemic of opiate addiction occurred in France, Prussia, or Austria, where soldiers had been treated with injectible morphine as often as had the Civil War veterans. It is not clear why American veterans, more so than those of other nations, became addicts. What is clear is that the major portion of the American opiate epidemic in the nineteenth century involved opium taken by mouth, not injected morphine. The source of the drug was primarily patent medicines.

Patent Medicine, an American Business

At the turn of the century the consumption of opium by Americans dwarfed that of the Europeans, and vast numbers of these consumers became opium addicts. In absolute numbers, by the beginning of the twentieth century the United

States had more opium addicts than it does today. This is remarkable when one considers that the population of the United States now is three times greater than it was in 1900.

Even more remarkable is the fact that the roughly 250,000 opium addicts in 1900 provided virtually no hazard to the general public. Being an opiate addict in those days was considered a minor character defect, very much like obesity in our time. By contrast, it has been estimated that today's heroin addicts account for as much as 50 percent of all crimes against property in the United States. The number of deaths related one way or another to the traffic in opiates and other illegal drugs may well exceed the number of violent deaths associated with any other form of crime. And this statistic does not take into account the deaths from overdose, which in New York City alone number in the thousands each year.

Why was opiate addiction so much more benign in the nineteenth century? What has changed so dramatically? The answer lies very much in the setting in which the drugs were used. It also relates to the way in which our legal system has regulated the use and abuse of opiates a hundred years ago and today.

Most of the opium consumed in America before the turn of the century was in the form of patent medicines. The growth of patent medicines derived in large part from the limitations of the medical profession. The number of physicians in those days was not adequate to deal with the rapidly expanding and geographically mobile American population. Also, compared with the standards of Europe, the quality of American medicine was woefully deficient. The major universities of England and the continent were training grounds for European physicians. In the United States, on the other hand, medical schools were mostly fly-by-night part-time operations in which young men would serve for a relatively brief time as apprentices to practicing physicians. Since prescription drugs were expensive and the quality of medical care offered by physicians left much to be desired, people naturally turned to the much cheaper patent medicines.

The patent-medicine industry in the United States was not regulated in the nineteenth century. Labeling was not required, so the public had no way of knowing whether one or another preparation contained opium or any other ingredient. Even many patent medicines that were advertised as "cures" for the opium habit frequently contained large amounts of opiates. This was certainly a valid way of dealing with withdrawal symptoms, but it was not exactly what customers expected when they purchased a "cure."

Since the manufacturer had free license to put almost any ingredient into his preparations, he would usually include opiates, often mixed with generous amounts of alcohol. These two ingredients ensured that there would be a genuine pharmacologic effect. At a minimum, the tranquilizing property of opiates would provide symptomatic relief for a rather broad range of medical complaints. Patients would feel that they were getting something for their money and would come back for more when the supply ran out.

The United States' patent-medicine industry grew almost exponentially in the second half of the nineteenth century, and with it the consumption of opium proliferated. The per capita importation of crude opium increased four- to fivefold from the 1840 figure of less than 12 grains per capita to more than 52 grains in 1890. A grain of opium would represent a fairly substantial dose. Incredibly, the quantity imported in 1890 comes to more than 50 doses for every man, woman, and child in the United States. Thus, it is not surprising that the United States was home to a quarter of a million opiate addicts in 1900.

The personal characteristics of the typical opiate addict of that day were greatly different from those of today's heroin addict, however. At the turn of the century the typical opiate addict was a middle-aged woman living an essentially normal life, married and raising a family. She purchased opium legally in the form of patent medicine and used it orally. Since she was fairly tolerant to the effects of the drug, her day-to-day activities could proceed much like her neighbors', with no evidence of physical or emotional disturbance. She was rarely worse off than being drowsy about midday and tending toward constipation. Because the effects of opium last longer when taken by mouth than when injected intravenously, she did not have sudden rushes of euphoria. By the same token, she rarely would experience withdrawal symptoms.

Most of these addicts to oral opiates had no adverse long-term medical effects. Opiate addiction in and of itself is not physically dangerous. Heroin addicts generally die of violent crime, hepatitis or AIDS from sharing needles, allergic reactions to materials mixed with the injected heroin, or simply overdose, which depresses the breathing process. Ingesting opiate extracts by mouth is safer than injecting morphine; but if dosage is reasonably well controlled, even opiates by injection are not inherently dangerous. There are many reported cases of opiate addiction among physicians, who have injected themselves regularly with pure, sterile morphine and continued to practice medicine productively, surviving in robust health to the age of 80 or 90.

The China Connection

Most of our present problems with opiate abuse stem from the massive efforts of the U.S. government early in this century to eradicate the essentially harmless addiction of a fraction of the population to the opium in patent medicines. The "cure" for opiate addiction turned out to be far worse than the disease. The forces that moved the federal government in this direction have a long history, one which goes back to the seventeenth century and to the imperialism of the British in China.

Opium reached China indirectly about the eleventh century from Arabian traders. However, it was used then only by a small portion of the population. In the 1600s there was some increased use of the drug. Chinese authorities, more alert to the dangers of opium than the Europeans, passed a law in 1729 banning the smoking of opium. Accordingly, the only way to obtain opium was by smuggling, and the major source was India. Smuggling at that time was becoming a big business, because the Chinese emperors for some years had prohibited the importation of all sorts of goods from Western countries, not just abusable drugs, in order to protect and stimulate Chinese industry. In the late 1700s the British East India Company, eager to penetrate the large and lucrative Chinese market for many products, obtained a monopoly on Indian opium and aggressively began to smuggle it into China. Actually, being upstanding British citizens, the company's executives took care not to be involved directly in something so illegal as smuggling. Instead, they would sell the opium from India to other middlemen, who in turn smuggled the opium into China.

The British were quite successful as dope dealers. In 1729, when the Chinese government first banned opium smoking, the number of chests of opium imported to China each year was 200. Within a decade after the British East India Company began its smuggling campaign, the number had increased to 5,000. By 1838, 25,000 chests were imported each year. The number of Chinese opiate addicts was clearly increasing at an extraordinary rate, despite efforts of the Chinese government to suppress use of the drug.

In 1820 the Chinese government issued an edict forbidding any vessel containing opium to enter the Canton River. In the next two decades, they enacted further restrictions and enforced them more and more stringently. Tensions heightened between the two countries to an intolerable level. Finally, in 1838, using as a pretext a drunken brawl between a few Chinese and British soldiers, the British commenced the famous opium wars against the Chinese. In less than a year the British army won a decisive victory. The spoils of victory were munificent. By the Treaty of Nanking, signed in 1842, the British were given the entire island of Hong Kong; they were paid 6 million dollars to reimburse the smugglers whose opium had been damaged or confiscated by the Chinese government; and they were awarded trade and residence rights worth far more than that. In 1858, at the end of yet a second opium war, Britain's access to Chinese markets became almost unlimited.

With such a felicitous end to the opium wars, England was free to pursue vigorously the exporting of opium from India to China. By the turn of this century, one-quarter of the Chinese population was addicted to opium. International outcries against the British were heard. The British themselves, commencing to feel somewhat guilty, entered into discussions to curtail the massive opium traffic. In 1909 an international conference on opium trade was convened by President Theodore Roosevelt. In part, no doubt, Roosevelt was acting out of moral outrage at the British. But part of his motivation was clearly U.S. self-interest, tinged perhaps with a bit of guilt.

In the late nineteenth century, large numbers of Chinese had come into the United States to work on the railroads, and many of these workers smoked opium. The myth quickly developed that opium came to the Western world from China; hence, opium addiction came to be considered a Chinese disease (when it wasn't being called the soldier's disease). The Chinese, who were competing with U.S. citizens for rights to jobs on the railroad, were regarded with derogation and blamed for much of what went wrong in society. The average American viewed the smoking of opium as a vice that went hand-in-hand with prostitution, sexual perversion, and crime—aberrations that many working-class Americans assumed to be innate features of Chinese culture. Americans generally failed to recognize that the opium the Chinese smoked was the same drug that so many red-blooded U.S. citizens were regularly ingesting in patent medicines.

In 1904 the U.S. Congress had capped off three decades of atrocious behavior toward Chinese immigrant workers by passing regulations excluding Chinese laborers from entering the country. In this racist atmosphere, travelers and immigrants to the United States from China found themselves victims of brutality at the hands of U.S. citizens. The Chinese government became increasingly incensed and threatened to take steps to prevent American financial interests from acquiring Chinese markets.

Whether his motivations were moral or commercial, President Roosevelt did exert leadership in negotiations to end the drug traffic into China. By 1912 international agreements mandating fairly strict regulation of trade in opium and other narcotics were being developed. At this point the U.S. government realized that we ourselves had no national restrictions on opium importation.

The Harrison Narcotic Act

The Harrison Act of 1914 did not explicitly make opium addiction illegal or forbid doctors from prescribing the drugs. However, a series of Supreme Court decisions interpreting this law progressively restricted the use of the drugs. Opiates could be prescribed to an addict only when he was within an institution and being specifically withdrawn from the drug.

The Pure Food and Drugs Act of 1906 already required that the ingredients of patent medicines be labeled. With the passage of the Harrison Act, opiate-containing preparations were gradually removed from the market. Middle-aged housewives, whose physical dependence usually was modest, somehow managed without the drugs. On the other hand, those addicts with a greater dependence on opiates were left with no legal recourse for maintaining their addiction. They were forced to turn to illegal drug sources. Hence, the modern underground narcotics industry was born. Illicit manufacturers, seeking a more compact product that was easier to conceal and smuggle, began to provide morphine for injection rather than crude opium for oral use. Since heroin is more than twice as potent as morphine per unit weight, it soon supplanted morphine. Of course, the cost of the illegal drugs skyrocketed to 50–100 times the price of the same drug from the no-longer-available legal sources. The addict was thus forced to turn to a life of crime to maintain his habit, and our modern heroin dilemma was born.

An important by-product of obtaining a chemically pure drug is that one can then modify the molecule and develop newer, potentially more effective or safer drugs. One of the easiest ways to modify a chemical with the structure of morphine is to add an acetyl group, which consists of two carbons and an oxygen. Adding two acetyl groups to morphine was accomplished by a group of chemists in the 1870s. The product, diacetylmorphine, was marketed in 1898 by the Bayer Company, two years after they introduced aspirin to the market. Diacetylmorphine was given the brand name "heroin" and was offered to the public as a cough medicine. Its major advantage was that it was thought to be nonaddicting.

How could the drug industry have been so mistaken about a 25-year-old drug? Everybody agrees that the standards demanded of a drug prior to its introduction to medical practice were far more lax in the nineteenth century than would be desired by even the most liberal drug-industry advocate today. But this cannot be the whole story. I believe we learn from this debacle something about the effects of opiates. The early twentieth-century physicians were not all fools. If heroin were so extraordinarily addictive that a few exposures would spell doom, then surely physicians would have required less than 25 years to pick this up. I believe that the heroin tale reinforces what many opiate pharmacologists know well, namely, that the ability of a drug to cause addiction depends as much or more on the social setting in which the drug is used and the mental set of the user as it does on the drug itself.

Heroin today is a degrading drug because of the way in which it is used—solely to secure a "high" by people who either feel hopeless about their life situation or are suffering from a severe conflict of identity. This quotation from a 16-year-old junkie living in New York's East Village in 1971 expresses the way in which addiction to opiates brings a focus of sorts to a bleak life. "The thing about heroin is that it gives a human being a purpose in life. It gives him an occupation, an identity, friends, a chance to be better at something and above all it takes up time."[4]

Unlike modern-day heroin junkies, patients treated with

heroin or morphine for legitimate reasons are more likely to feel uncomfortable and nauseous than to feel euphoric. If the drug is used only on a limited basis, for the treatment of a specific symptom, severe physical dependence does not necessarily eventuate. Indeed, the whole history of opiates through the ages teaches us that what matters is not so much the drug itself as how people use it and how society views this use.

· 3 ·

Stalking the Opiate Receptor

In the summer of 1972, as Candace Pert and I began our search for the opiate receptor, what evidence did we and other scientists have that opiate receptors even exist? To understand this, we should first review a little bit about how hormones and neurotransmitters mediate the "dialogue" among different organs of the body, as well as between neurons and adjacent neurons, glands, or muscles.

Information transfer is a crucial aspect of body function. Hormones provide one of the most dramatic examples of pinpoint precision in communicating across seemingly vast distances. The pituitary gland, located at the base of the brain, releases hormones that regulate activities of the adrenal glands, thyroid gland, and gonads. These other glands in turn release hormones that act throughout the body. For instance, the two adrenal glands (located above the kidneys) release cortisol, which influences the brain, liver, lungs, and muscles. Insulin, released by the pancreas into the blood stream, influences most cells of the body, enabling them to accumulate glucose from the blood. Glucose (blood sugar) is the primary source of the cell's nourishment; without insulin, cells quickly become "starved," cease to function properly, and eventually die.

The amounts of these hormones manufactured and released into the blood are almost infinitesimally small. Consider that the blood level of insulin is normally less than 1/10 millionth of a gram per liter and there are 454 grams in a pound. How could insulin's target cells detect such low concentrations? Moreover, how does insulin know which cells are valid targets?

The brain presents an even more complex riddle, with its 10 to 50 billion separate nerve cells, or neurons. All of our ability to think and feel stems from the ways in which these neurons communicate with one another. They do this by releasing neurotransmitters, chemical messengers liberated from nerve endings when the neurons fire (see figure 2). These

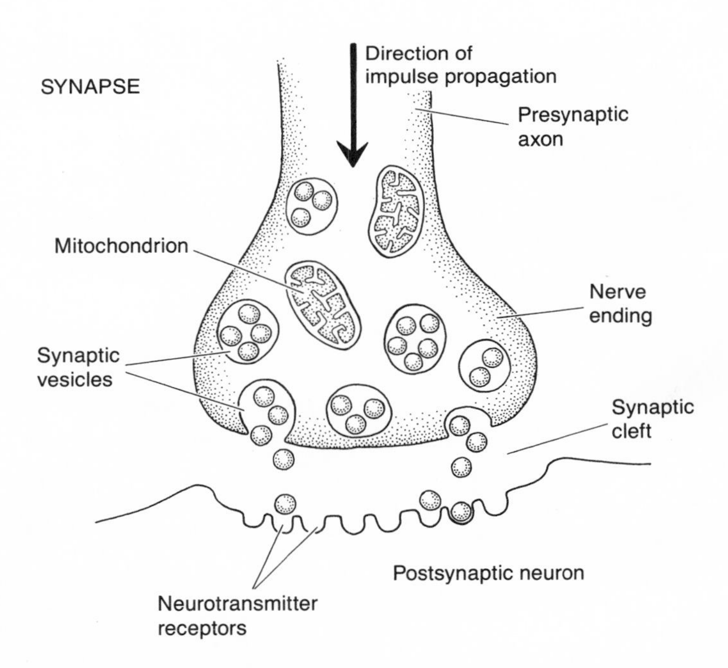

Figure 2. Neurotransmitter molecules (depicted here as small balls) are stored in synaptic vesicles inside nerve endings. In response to an electrical impulse in the presynaptic axon, these vesicles fuse with the membrane of the nerve ending, releasing neurotransmitter molecules into the space between neurons (synaptic cleft). The neurotransmitter molecules diffuse across the space to the postsynaptic neuron and bind to receptors on its dendrites, thereby altering the activity of the neuron. Mitochondria—energy-generating organelles in cells—fuel the process of neurotransmitter release.

messengers travel across the tiny spaces, at synapses, that separate nerve endings from the receiving elements (dendrites) of adjacent cells. Each neuron may possess several thousand nerve endings, which branch off of the nerve's long axon. Each of these nerve endings is just a synapse away from a dendrite of an adjacent neuron. A given neuron may possess many dendrites, each of which receives input from many nerve endings. For the sake of precision, a neuron would not want to influence every adjacent neuron, so the challenge is how to discriminate between those to be communicated with and those to be avoided.

Receptors provide the solution to this dilemma, both for hormones and for neurotransmitters. Receptors are cell proteins that recognize specific hormones or neurotransmitters with precision and fidelity; the chemical "fit" of a hormone or neurotransmitter and its receptor is much like the fit of a key in a lock. The perfect match whereby an insulin receptor recognizes insulin permits detection of the low circulating concentrations of insulin. Since only cells with insulin receptors recognize and respond to insulin, these are the target cells of the hormone. The same story holds true for neurotransmitters. Only those neurons with receptors for a given neurotransmitter can respond to it.

Many important drugs in medicine have been designed to mimic neurotransmitters or hormones; other drugs have been designed to block the effects of endogenous neurotransmitters or hormones. Such drugs, almost by definition, act at receptor sites. But many other drugs were developed without any knowledge of the particular way in which they influence the body. Some drugs do not seem to act by specific receptor-like recognition sites. For instance, local anesthetics, like Novocain, block the firing of sensory nerves merely by dissolving in the fatty lining of the nerves and interfering with the passage of nerve impulses. Other drugs exert such selective effects that one can reasonably expect that they act via receptors. For instance, a drug that acts in extremely low doses is likely to use receptors. Similarly, any drug that influences only a few organs of the body probably binds to receptors located only on those organs. And a drug that loses all of its effects following a minor change in its molecular structure probably acts through a receptor recognition site so selective that it cannot tolerate much of a change in the drug's chemical makeup.

Emergence of the Receptor Concept

It is hard to pinpoint an exact date when the receptor concept emerged; it seems to have percolated into scientific awareness gradually at the beginning of the twentieth century. Many pharmacologists attribute the notion of receptors to the renowned scientist Paul Ehrlich, who introduced this way of thinking when he developed arsenic-related chemicals as the first "antibiotics" to treat syphilis. Ehrlich found that small changes in the structure of his "magic bullet" drug, salvarsan, caused it to lose activity. He postulated that the drug must be acting through a "lock-and-key" recognition site, and even drew diagrams depicting such interactions.

Translating Ehrlich's diagrams from conjecture into reliable experimental identification of receptors as biochemical entities took many more years. The first approaches occurred in the early 1960s, when Dr. Elwood Jensen at the University of Chicago injected radioactively labeled estrogen into rats and found that it concentrated selectively in organs, such as the uterus, that are the known targets of the hormone.

The work of Jensen and his colleagues with estrogens, extended by others to the male sex hormones (androgens) and cortisol, the adrenal hormone, represents the first generation of molecular receptor research. These receptors are found either in the cytoplasm of cells—that is, in the parts of the cell that are outside the nucleus but inside the cell's outer membranes—or on the membrane that encloses the nucleus. These receptors cannot explain the actions of hormones such as insulin or the numerous hormones released by the pituitary gland, all of which are peptides (small proteins). Peptides and larger proteins generally cannot diffuse through membranes of cells to reach the cytoplasm or nucleus. If they are to act through receptors, such receptors must be located on the outside surface of the cell's membrane.

Similarly, neurotransmitters cannot act through intracellular receptors. Today we know of about fifty neurotransmitters. Some are peptides, others are amino acids, while still others are amines—nitrogen-containing molecules derived from amino acids. All of the known neurotransmitter molecules possess electrical charges that would impair their ability to diffuse into cells. Hence, they too must bind to receptors on the outside of cell membranes. Moreover, the extremely rapid transmission of information from neuron to neuron cannot wait for a neurotransmitter to penetrate the interior of a cell before reaching its receptor.

The first well-characterized cell membrane receptor was for insulin. In 1970 Drs. Pedro Cuatrecasas and Jesse Roth at the National Institutes of Health in Bethesda, Maryland, identified insulin receptors by a fairly straightforward approach, measuring the binding of radioactive insulin to membranes of cells that were known targets of insulin. Our identification of opiate receptors used techniques similar to theirs. Indeed, arguments for the existence of opiate receptors resemble those marshalled as evidence for insulin receptors.

Potency

The first piece of evidence for opiate receptors has to do with potency. Some opiates act in very small doses associated with low concentrations in the blood. Etorphine is an opiate drug that is ten thousand times more potent than morphine in relieving pain. It is also exquisitely precise; very small changes in the chemical structure of etorphine can reduce its potency by a factor of a thousand. It is hard to imagine how a drug could be recognized by target cells in such a low concentration and be so sensitive to small changes in chemical structure unless it is bound to highly selective receptors.

Evidence from Antagonists

A second piece of strong evidence in favor of the opiate receptor was the ability of certain drugs, at very low doses, to block the action of opiates.

A drug which, after binding to a cell, alters cellular activity is called an agonist. By contrast, other substances can be bound just as selectively by a cell but, for reasons we still do not understand, fail to have any further effect on cellular function. These substances are called antagonists. Something about the structure of an antagonist molecule seems to throw a monkey wrench into the operations of the cell. The situation is a little reminiscent of the small boy who can fit into the driver's seat of a car but who does not know how to turn on the ignition switch.

Naloxone was the first widely used "pure" opiate antagonist. It is pure in the sense that, unlike morphine, it has no capacity to produce euphoria or to relieve pain. But if you inject an animal first with a very low dose of naloxone and later with a hefty dose of morphine, morphine will not produce analgesia. And if an animal has already been injected with morphine, an injection of a minute amount of naloxone will undo the effects of the opiate. One cannot easily explain the ability of such tiny concentrations of naloxone to prevent or undo the effects of morphine without invoking the existence of specific opiate receptors. Naloxone seems to have a very high affinity for these sites on the cell surface and can find them at very low doses. It seems to sit on the opiate receptor without producing any of the cellular changes necessary for analgesia, but by occupying the receptor, naloxone blocks the access of conventional opiate agonists such as morphine or heroin (see figure 3). Furthermore, naloxone seems to have such a high affinity for opiate receptors that it can knock an opiate off a receptor in order to occupy the receptor itself.

This fact has made naloxone a lifesaving drug in cases of opiate overdose. Opiates do not kill by damaging the brain

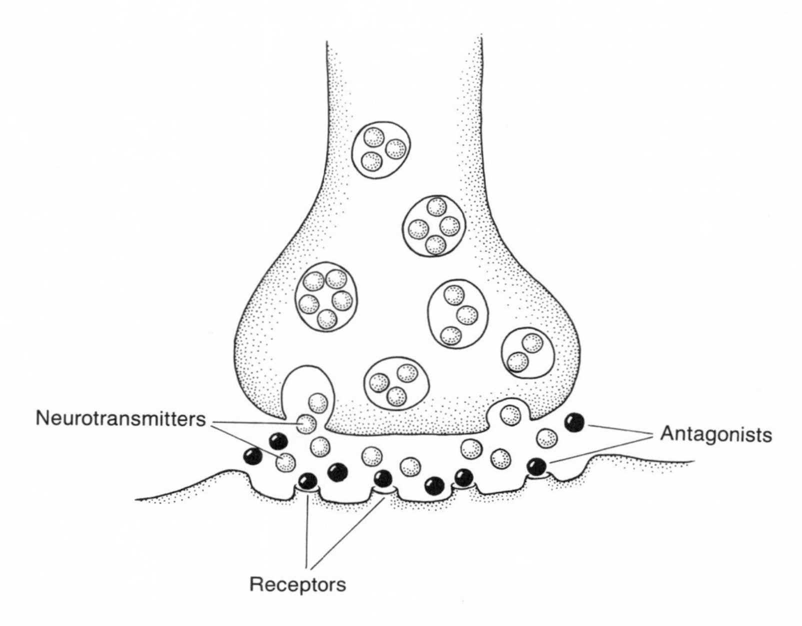

Figure 3. Antagonist drugs (shown here as small black balls) fit into neurotransmitter receptors in the same way that neurotransmitter molecules do. However, unlike neurotransmitters, the antagonists do not activate the receptor. Instead, their presence prevents neurotransmitter molecules, or drug molecules which mimic them, from gaining access to the receptor.

directly. Rather, their major lethal effect is to depress the rate of breathing so severely that the patient lapses into coma from lack of oxygen to the brain. Most people breathe about 15–20 times a minute, but patients with opiate overdoses may breathe only 2 or 3 times a minute. Injecting these patients with a minute dose of naloxone almost miraculously reverses this depressed respiration rate. Within 10 seconds of an intravenous naloxone injection, formerly comatose patients will sit up, totally alert and almost normal.

Today, there are many other antagonist drugs besides opiate antagonists that play a major role in clinical medicine. Two principal groups of drugs in cardiology block receptors for the neurotransmitter norepinephrine. Norepinephrine is a neurotransmitter in the brain, but it also acts on sympathetic nerves throughout the body that control involuntary processes such as beating of the heart and constriction of blood vessels. Two major types of receptors for norepinephrine have been identified, designated alpha and beta. When norepinephrine acts at beta receptors in the heart, it speeds the heart rate. For a patient with angina, a rapidly beating heart uses up oxygen, and the resultant oxygen deprivation triggers the characteristic pain of angina. Beta receptor antagonist drugs (beta blockers) slow down heart rate and thus relieve the symptoms of angina; they are, consequently, among the most widely used drugs in clinical medicine. For developing the first beta blockers and characterizing this type of norepinephrine receptor, Sir James Black shared the 1988 Nobel Prize in Physiology and Medicine. At alpha receptor sites, norepinephrine causes constriction of blood vessels, which raises blood pressure. Alpha antagonists block these effects and thus reduce the blood pressure of patients with hypertension.

If we knew why an opiate agonist such as morphine is an agonist and why an antagonist such as naloxone is an antagonist, we would probably be close to solving basic riddles about how cells in the nervous system communicate with one another. Exactly how the recognition of a neurotransmitter molecule or an agonist drug brings about a change in cellular activity, while the recognition of an antagonist does not, is still a mystery. It is one of the central questions of neurobiology, much as the genetic code was for molecular biology. Most neurobiologists believe that this change in the cell's activity is brought about by a second part of the receptor, one that is different from the recognition site. Together, these two parts make up the "receptor complex." But isolation of the full receptor complex and insight into how it operates is a field in scientific infancy. Recognition sites, on the other hand, have been isolated for many neurotransmitters and drugs, including the sites that recognize opiates. When most researchers use the term "receptor," we are referring only to the recognition site, not to the entire receptor complex.

Mirror Images

A final piece of evidence in support of opiate receptors grew from the understanding that many chemicals exist in mirror-image forms—two chemically identical versions of a molecule that differ only in that one is "left-handed" and the other is "right-handed," much like a pair of gloves. Mirror-image forms of chemicals are called optical isomers because they differ in the way they rotate a plane of polarized light. Louis Pasteur, more famous for his germ theory of disease, pasteurization, and rabies vaccination, first characterized optical isomers.

Two mirror-image forms of a drug may also differ markedly in their ability to cause a change in the body. This phenomenon is called stereospecificity. For the opiate drugs, stereospecifity is the rule. In general, the form of an opiate that rotates light to the left (called levorotatory, from the Latin *levo,* left) is active in the body; the right-handed (*dextro*) version is not. Yet the chemical constituents of the two forms are identical. This argues that opiates must be acting at a site so selective that it can bind the left-handed version of a drug while passing up its chemically identical right-handed version. So selective a recognition site is, by definition, a receptor.

Binding Experiments

What kinds of experiments could one conduct with opiates to isolate the opiate receptor? In modern biochemistry, when one wants to find out where a drug is exerting its action in the body, one uses radioactive chemical tracers. The advent of this technique in the 1950s transformed the plodding science of biochemistry into the fast-moving field it is today.

The first step in tracing a drug is to replace one of the atoms of the drug with a radioactive isotope of the same element. Next, the tissue being tested is homogenized and then mixed in a test tube with the radioactively labeled drug. Homogenizing the tissue breaks down all the barriers between the outside and inside of cells so that, wherever the receptor for the drug may be in the tissue, it will have access to the radioactive drug. After the drug and the tissue sit together for a reasonable time, one can then separate the radioactive drug bound to the tissue from the unbound drug. The simplest way to separate the two is to filter. One merely pours the tissue-plus-drug through a filter maintained under vacuum pressure. The tissue, and whatever amount of drug is bound to it, will be trapped by the filter, while the unbound drug in solution will be washed through. We then use a special machine, much like a geiger counter, to count the radioactivity on the filters.

It all sounds very simple. Indeed, the very simplicity of the binding approach had prompted numerous researchers since the mid-1950s to attempt to measure the binding of radioactive opiates to brain membranes of experimental animals. They did indeed observe binding of opiates, but most of this binding was occurring at sites in the brain that were not the true opiate receptors. Two pieces of evidence led to that conclusion. First, with this technique, drugs that were known to be very weak as painkillers would bind just as readily to brain membranes as drugs that were known to be powerful analgesics. Second, the two optical isomers of opiates would bind equally well with this technique, a fact that does not fit at all with the stereospecificity of opiate drugs.

Why so many researchers failed to identify the true opiate receptor is not hard to discern. The major reason is that there are so few opiate receptors in the brain. One can make an educated guess as to the actual number, if one knows the effective dose of an opiate and its concentration in the brain, and if one makes certain assumptions about the size of the opiate receptor. Vincent Dole, an eminent biochemist famous for developing the methadone maintenance treatment of heroin addicts, published such a calculation in 1970. He estimated that opiate receptors account for only about one-millionth by weight of the brain.

A second reason for the failure is the fact that opiates bind so readily to many brain constituents that have nothing to do with opiate receptors. Being relatively simple molecules, opiates have portions which are electrically charged and other portions which are not charged but are chemically "sticky." Brain membranes contain all sorts of proteins, lipids, and carbohydrates, some of which are highly charged and others of which lack electrical charges but have "sticky" properties complementary to the stickiness of opiate drugs. The positive charges on an opiate molecule would be expected to bind to negative charges on the many sugars and proteins of brain membranes. The uncharged portions of opiate drugs will very likely adhere to uncharged parts of membrane protein and lipids, much as chewing gum sticks to the soles of shoes. If one considers that there is an almost infinite number of charged and uncharged molecules all over the surface of brain membranes that can bind nonspecifically to opiate drugs, how could one ever hope to pick out the radioactive signal of an opiate bound to the tiny number of true receptors, above the noise of all these other meaningless chemical interactions?

Goldstein's Forays

In 1971 Avram Goldstein, Professor of Pharmacology at Stanford, published a paper in which he explained all of these problems quite clearly and described some experiments he had conducted that hinted at progress toward identifying opiate receptors.[1] At the Gordon Conference I had attended back in 1971 I first heard him discuss these issues. I then read his paper and began thinking about the problems he had run into and some possible solutions.

In Goldstein's experiments he hoped to take advantage of the stereospecificity of opiates to ascertain how much binding to brain membranes involved true opiate receptors and how much involved other sites. Here is what Goldstein did: He mixed radioactively labeled levorphanol, an opiate drug, together with membranes from a mouse's brain and counted the radioactivity that was bound. Next he mixed radioactively labeled levorphanol with nonradioactive levorphanol, put these in a test tube with mouse brain, and then counted the amount of radioactivity that was bound. Finally, he mixed the radioactively labeled levorphanol with nonradioactive dextrorphan (the optical isomer of levorphanol), combined these with mouse brain, and counted the radioactivity that was bound. Goldstein reasoned that if all of the binding of radioactive levorphanol that he counted in his first step was occurring at true opiate receptors, the nonradioactive levorphanol added in the second step would compete with the radioactively labeled levorphanol for the receptor sites; therefore the counts of radioactivity in the second step should be lower than in the first. Dextrorphan, on the other hand, because it is the pharmacologically inactive one of the pair of optical isomers, would not compete with the radioactively labeled levorphanol at the binding sites, and so the count in the third experiment would be approximately the same as the count in the first.

The results did not work out like that. In most cases dextrorphan reduced the amount of radioactive levorphanol bound as much as did nonradioactive levorphanol. Goldstein repeated the experiment again and again. With many replications he noticed a very slight tendency for levorphanol to reduce the number of counts more than dextrorphan. It seemed that about 2 percent of the total of radioactivity bound to the membranes was inhibited more by levorphanol than by dextrorphan. He suggested that the small 2 percent difference reflected a minute signal of radioactive levorphanol binding to opiate receptors above the vast, noisy background of radioactive levorphanol binding to all sorts of nonspecific sites.

Though other researchers before and after have discussed the conditions that must be met before binding of a drug to membranes can be assumed to represent a biologically relevant receptor, none of them had written as thoughtfully about these problems in connection with opiate receptors as did Goldstein. The 2 percent of stereospecific binding he detected suggested that there might be a measurable opiate receptor, if only one could develop stronger tools to pull it out. Over the next three years Goldstein was able to amplify the signal and clarify the nature of the material that was binding his radioactive levorphanol. Alas, it turned out that the binding did not involve the opiate receptor after all. Instead, Goldstein had detected binding of radioactive levorphanol to a brain lipid that has the interesting capacity to discriminate between the optical isomers of opiates.

Despite the fact that he had not identified the opiate receptor, Goldstein's publication in 1971 did influence thinking in the field and certainly affected my own conceptions about receptors. I read through his paper again and again in the months following that Gordon Conference, trying to understand why he had run into problems and to identify a few technical "tricks" that might boost the extent of receptor binding of the radioactive opiate.

My reasoning went this way: One can assume that the nonspecific binding sites would have much less affinity for opiates than would true opiate receptors. Furthermore, a receptor with high affinity for a drug should be able to bind an extremely low concentration of the drug far more efficiently than a site that has a low affinity for the drug. If so, then we should use the lowest possible concentration of a radioactive opiate in our binding experiments. Of course, if one uses too little radioactive drug, then the counting machine will not be able to measure it. Accordingly, what we should use is a radioactive drug with the greatest amount of radioactivity possible per molecule of drug. With a high level of radioactivity per molecule, we could add very few molecules and still have a reasonable amount of radioactivity to detect. The levorphanol that Goldstein used possessed a rather low amount of radioactivity per molecule, so that he was forced to add high concentrations of the radioactive drug. The result was binding to all sorts of sites, very few of which were the opiate receptor. I remember writing down comments in the margins of Goldstein's article to the effect that "we better get some highly radioactive opiate custom made."

Another trick that I thought might help in detecting receptor binding would be to wash the membranes after binding the radioactive drug to them. If opiate receptors bind the drug more tightly than do nonspecific sites, then one should be able to wash away the radioactive drug bound to nonspecific sites without washing away the drug bound to opiate receptors. If one used filter paper to trap the radioactive drug bound to brain membranes, then it should be easy to pour wash water over the filter and suck it away rapidly with a vacuum. Goldstein had not used a filtering apparatus in his experiments. Instead, he had simply centrifuged the mixture of radioactive drug and brain membranes. A centrifuge spins tubes containing the mixture of tissue, water, and drug so rapidly that membranes with the radioactive drug bound to them concentrate as a pellet at the bottom of the tube. In contrast with a filtration experiment, with a centrifuge there is no way to wash off nonspecifically bound radioactive opiate from the pellet in the centrifuge. I made another note in the margin of Goldstein's paper which said, "Let's try filtration."

Clues from Insulin

Nothing happens in a void. The specific experimental strategy that I thought might succeed in isolating the opiate receptor derived in good part from my friendship with Pedro Cuatrecasas, who had moved from NIH to Johns Hopkins and now happened to occupy the laboratory next to mine. In 1970 Cuatrecasas had used the same approach to identify the insulin receptor. First he prepared insulin with very high amounts of radioactivity per molecule, and then he measured its binding to fat and liver membranes using a filtration apparatus. Washing the membranes thoroughly but extremely rapidly, he could remove nonspecific binding without affecting insulin interactions with the receptor.

As in many other instances in science, discovery of the insulin receptor by binding techniques was accomplished almost at the same time independently by Dr. Jesse Roth and his colleagues, also at the National Institutes of Health. Over the next few years Cuatrecasas and Roth used these simple, sensitive binding techniques to elucidate many of the mysteries of how insulin regulates blood sugar and other metabolic functions. One common type of diabetes (called type-1 or insulin-dependent diabetes, or, formerly, juvenile diabetes) was known to develop when the pancreas loses its ability to produce enough insulin. But Cuatrecasas and Roth were able to show that abnormalities in the insulin receptor itself which interfere with its ability to respond to insulin—rather than a shortage of insulin in the blood stream—sometimes occur in a second very common type of diabetes (which is called type-2 or non-insulin-dependent diabetes, or, sometimes, adult-onset diabetes because, unlike type-1, it usually first appears in adulthood).

At first my interest in Cuatrecasas' work was largely confined to admiring the brilliance of his attack on one of the central questions of medicine, how insulin acts in diabetes. Then one day in early 1972 I read a publication in *Science* magazine by a group of researchers at Washington University in St. Louis. These scientists had worked out the amino-acid sequence of the protein called nerve growth factor. Nerve growth factor regulates the growth of certain populations of nerves in the body and is one of a presumably large number of growth factors that determine the development of all tissues of the body. The amino-acid sequence of nerve growth factor showed many similarities to the sequence of insulin. (All proteins are composed of strings of amino acids, of which there are twenty altogether. The particular order of these "beads on a string," and the way the strings fold back on one another, determine the protein's action in the body. See figure 4.) I suspected that the techniques which had been successful in identifying insulin receptors might be applied to studies of the receptor sites for nerve growth factor. A recently arrived postdoctoral fellow in my laboratory, Shailesh Banerjee, was in search of a project. I asked Cuatrecasas if he might collaborate with Banerjee and me in a search for the nerve growth factor receptor. Cuatrecasas was enthusiastic, and we set to work right away. During the first months of 1972 I was able to assimilate in the best way possible—namely, to "learn by doing"—all the techniques and thinking strategies that go into isolating receptors. The project was eminently successful, and over the course of the next few years we learned a great deal about how nerve growth factor interacts with specific

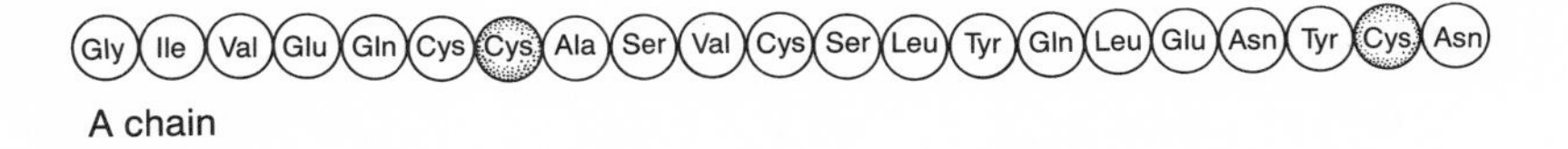

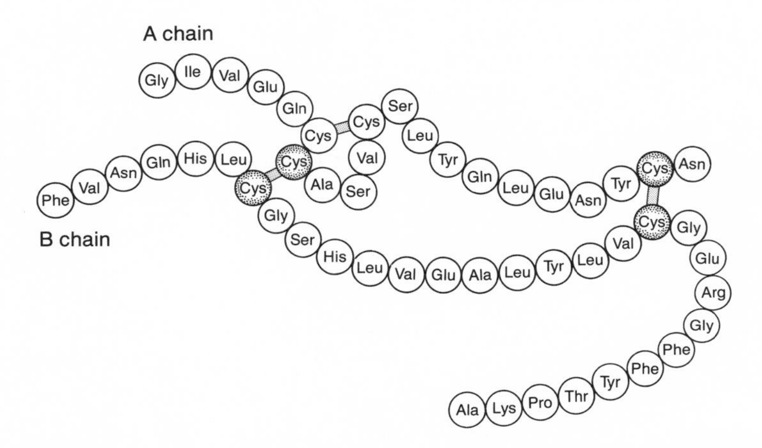

Figure 4. Proteins are long chains of amino acids (top) which assume a three-dimensional shape by folding and sometimes, as in the example of human insulin shown here, by combining with other chains. The particular sequence of amino acids dictates how a chain will fold and consequently what the protein's overall shape, surface features, and chemical activity will be.

binding sites to direct the growth of embryonic nerves to their adult stage.

Nevertheless, in the fall of 1972, when Candace and I began the search for the opiate receptor, success with measuring receptors by binding techniques had been confined largely to hormonal receptors. While in theory opiate receptors ought to exist, no one seriously believed that you could detect them with the binding procedures that worked so well with hormones. Indeed, Avram Goldstein's failure argued to most people that the task would be virtually or absolutely impossible. But Goldstein had not tried Cuatrecasas' filtration and washing technique, nor had he used a highly radioactive tracer. Candace Pert and I commenced to do just that.

First Failures

In her just-completed research measuring the accumulation of choline by intestinal strips, Candace had used a homemade filtration device our departmental workshop had fabricated some years earlier to measure neurotransmitter accumulation in brain tissue. We knew that opiates act upon the intestine as potently as upon the brain. Accordingly, we used intestinal strips instead of brain tissue as our source of potential opiate receptors.

What about the all-important highly radioactive drug? The New England Nuclear Corporation is the principal company that produces radioactive chemicals for use by medical researchers. Because radioactive chemicals are so critical to science, researchers often peruse the New England Nuclear catalogue and develop experimental ideas simply on the basis of what is available at the store. In 1972, New England Nuclear started listing as a readily available catalogue item the opiate dihydromorphine labeled with the radioactive isotope tritium. Tritium, designated with the prefix ^{3}H-, is the most widely used and convenient isotope for biomedical research. Not only was the price of radioactive dihydromorphine fairly reasonable, but it possessed almost one thousand times more radioactivity per molecule than did the levorphanol Avram Goldstein had employed. Accordingly, one could add to the test tube only a tenth of a percent as much radioactive dihydromorphine as Goldstein had added of radioactive levorphanol. This would surely make it more likely that the drug would seek and bind to opiate receptors rather than to nonspecific binding sites.

Armed with our new tools, Candace mixed radioactive dihydromorphine together with intestinal strips and a solution of salts resembling the body's internal fluids. After allowing the mixture to incubate for awhile, she filtered it. With our relatively crude filtration apparatus it was still possible to wash the intestinal strips a few times with the buffer solution. In some beakers we had only radioactive dihydromorphine together with the intestine and buffer solution, while in other beakers we included either nonradioactive levorphanol or nonradioactive dextrorphan to check for the stereospecificity of binding, as Goldstein had done. To our chagrin, we failed to detect any stereospecific binding. Candace tried valiantly to modify the experimental conditions. She checked out a variety of temperatures. She incubated for varying durations of time. She modified the acidity of a buffer solution. She tried washing different numbers of time. All to no avail.

What neither of us realized then was that dihydromorphine is exceedingly sensitive to light. Like most modern laboratories, ours is drenched in light from an imposing array, row upon row, of large, intense, white fluorescent light bulbs. The dihydromorphine had little chance of surviving the onslaught of destructive visible and ultraviolet rays.

All of this occupied the better part of a month. When a student has been working on a pilot experiment for more than two weeks, I become increasingly edgy. Patience has never been one of my virtues. And with some justification. Too much patience can be disastrous for a scientist. I have witnessed students immersing themselves for several years in dead-end projects and emerging at the end of this time without a suitable PhD dissertation. Such a horror story has never befallen any of my students, because I take great pains to ensure that dead-end streets are quickly abandoned.

Thus Candace and I were confronted with the dilemma: Should we quit before we wasted even more time and money, or should we keep trying? I was feeling sufficiently discouraged that dropping the opiate receptor work seemed the best choice. Candace, on the other hand, felt more inclined to give it another try. Dihydromorphine, a close relative of morphine, is an opiate agonist. Perhaps receptors would be detectable only with antagonists. It is conceivable that antagonists would have a higher affinity for the opiate receptor than agonists. We knew that naloxone is clinically effective in much lower doses than is morphine. Because of its great potency, naloxone seemed like an ideal candidate for labeling opiate receptors. Unfortunately, naloxone was not available in a radioactive form. The only radioactive opiate one could purchase directly from a catalog was dihydromorphine. To custom-label naloxone with radioactivity would be substantially more expensive than simply buying a stock item from a catalog.

I inquired of the New England Nuclear Company about costs. They quoted some rather large figure for preparing radioactive naloxone, which they would guarantee as being stable and of 95 percent or greater purity. I do not recall the exact dollar figure, but it seemed gargantuan to me and certainly out of the question. For poor folk like us, New England Nuclear was willing to offer an alternative, however. They would label the naloxone with tritium and send it to us in a crude, unpurified state with no guarantees whatsoever. It was up to us to purify the material. The price would be $300, even if we failed to recover any usable radioactivity. I agreed to accept the risks.

Breakthrough

It turned out to be rather easy to purify the substance. Best of all, the radioactive naloxone was quite stable and simple to use. I recall the day of Candace's first experiment. The setup for the experiment, including filtration and washing, was identical to that for dihydromorphine. Roughly 1,500 counts bound to the intestine when Candace added radioactive naloxone alone. In the presence of nonradioactive levorphanol, the number of counts fell to 600, indicating that the levorphanol had competed with naloxone at the binding sites. When she added dextrorphan, the number of counts bound was identical to when she used radioactive naloxone alone. Here was stereospecific binding. The amount of binding was robust. Even in these first experiments it was clear that we had enough naloxone bound to the membranes in a stereospecific fashion that we could conduct experiments rapidly and accurately.

The experiments necessary to pin down our finding—to determine for sure whether we were dealing with the true opiate receptor—were relatively straightforward. The existence of stereospecific binding is a necessary but by no means sufficient condition for ensuring that one is monitoring the opiate receptor. As we had seen in Goldstein's studies, even an extremely ubiquitous lipid available commercially from chemical supply companies can bind opiates in a stereospecific fashion.

We immediately proceeded to screen a series of opiate drugs to see if their relative potencies at the binding sites we had isolated fit with their pharmacological potencies. We obtained a very convincing correspondence. Even apparent exceptions were the sort that prove the rule. For instance, codeine is a fairly potent opiate, roughly 15–20 percent as potent as morphine itself. However, we found that codeine had negligible affinity for the naloxone binding sites. This discrepancy worried us at first, until we checked the pharmacologic literature. Codeine appears not to have any action by itself. Instead, codeine is first converted by enzymes in the liver to morphine, which then passes to the brain to cause pain relief. In other words, the chemical entity codeine is not an active opiate. It is merely a precursor drug for morphine.

We were pleased to find that the relative potencies of opiates at the binding sites matched their pharmacological actions. What about nonopiates? Would aspirin bind to the sites we had isolated? What about the wide range of psychotropic drugs? With our simple, sensitive, and readily performed technique, we were able to evaluate over 50 drugs in short order. No drugs other than opiates bound readily to the binding sites we had detected with radioactive naloxone. Clearly we were dealing with a very specific and selective opiate receptor.

Our first experiments had used intestinal strips. We moved rapidly to brain tissue, using a more modern filtration apparatus that had worked well in the insulin and nerve growth factor research. To our delight, in the brain we discovered not only opiate receptors but marked differences in their number in different parts of the brain. Presumably opiates would act most prominently in those areas of the brain that contain higher densities of receptors. Thus in these first experiments in the brain, we were witnessing clues that might lead us to understand how opiates influence some brain functions, such as pain perception and mood, much more than they influence others.

· 4 ·

Staking the Claim and Taking the Blame

Julius Axelrod often admonished me, "Sol, when you have made some discovery or think you have solved a problem, never stop there. Every discovery raises more questions than it answers." How well this advice applied to the opiate receptor! We were excited that what had seemed an almost impossible task had succeeded. It was not like many scientific projects, in which you scratch your head and agitate back and forth in your mind over just what a given experiment signifies. To our eyes there seemed to be no ambiguity. This was the genuine, bona fide opiate receptor, the site responsible for the extraordinary things that opiate drugs do to human beings.

But euphoric as we were in this discovery, Axelrod's advice rang forth again and again in my mind. Identifying opiate receptors was merely the tip of the iceberg. Our simple and sensitive technique to count opiate receptors provided us with a powerful molecular tool with which to probe the really tantalizing question that was on the minds of many Americans, namely, what is the nature of opiate addiction? The escalating heroin epidemic in American ghettos and the heroin addiction of American soldiers in Vietnam were being trumpeted in the newspapers every day. If it were possible to understand the nature of opiate addiction, perhaps one might be able to devise some means of combating it effectively.

Opiate addiction was not the only avenue possibly opened up by this research. People become dependent on a large number of substances—alcohol, nicotine, barbiturates, tranquilizers such as Valium, all sorts of sleeping medications, cocaine, amphetamines—the list could go on and on. In all cases, addiction displays certain common properties. First there is tolerance: after using a substance chronically one requires higher and higher doses to produce effects obtained with far lower doses when the person was first exposed to the substance. One also becomes physically dependent on addictive substances: that is, stopping the substance leads to withdrawal symptoms, which are usually the opposite of the initial effects of the drugs themselves. Alcohol depresses our mental functions; in alcohol withdrawal, referred to in its worst state as delirium tremens, victims enter a state of hyperexcitability and even experience convulsions. Amphetamines are stimulants; during amphetamine withdrawal, addicts become extremely somnolent and will sleep for days on end.

Besides tolerance and physical withdrawal, addiction is associated with a compulsive drug-seeking behavior. Simply put, an addict will relentlessly seek out his drug, regardless of whether or not he is experiencing any withdrawal symptoms. The drug-seeking behavior seems to be totally independent of the presence of tolerance or physical withdrawal. It has been the bane of physicians attempting to curb opiate addiction. For years, heroin addicts from the streets of Harlem have been shipped off to the Federal Narcotic Hospital in Lexington, Kentucky, where they are gradually withdrawn from the drug. In the closely guarded environment of what is essentially a Federal prison, the former addicts are maintained completely free of all drugs for many months. At the end of this time their tolerance to heroin has abated, and one would think that they are essentially cured. However, within days of returning to their old haunts in New York, they find themselves in the company of their neighborhood pusher, and in a few weeks are as severely addicted as ever before.

Some authorities (including myself) argue that compulsive drug seeking is sociologic. They maintain that there is nothing inherently biological that makes addicts return to their drug. Instead, the culprit is mental set and physical setting, the emotional pressure of old friends, the absence of nondrug alternatives for interesting living, the sheer emptiness of ghetto life. The subsequent experience of American soldier-addicts in Vietnam seems to lend support to this view. Upon return to the United States, very few became readdicted. Perhaps if they were thrust back into an environment similar to Vietnam, these soldiers would again become opiate addicts.

On the other hand, some researchers feel that physical, biological factors account for the extraordinary relentless craving that addicts have for their drug. Psychologists have demonstrated such long-term craving in rats and other experimental animals.

Whatever the source of drug craving, everyone agrees that the triad of tolerance, withdrawal, and drug-seeking behavior characterizes all the major addictions. Because of this formal similarity of the addictive processes for different drugs, perhaps the same or closely related fundamental mechanisms underlie all addiction. Thus, if we could find a molecular mechanism that accounts for addiction to opiates, a very similar mechanism might explain addiction to other substances.

Countless other questions begged for our experimental ministrations. Do people become addicted because of a change in the number or properties of opiate receptors? What happens biochemically in neurons after opiate receptors have been activated? How do opiate receptors tell the difference between agonists and antagonists? How does recognition of opiates at their receptor sites elicit the unique changes in pain perception and emotion that has made these drugs so valuable as medicines and so vulnerable to abuse? And at the most basic level, why does so complex an apparatus as the opiate receptor exist? Was it created merely to tempt humankind, like the mythical apple proffered by the serpent to Eve? Or was there a biological advantage that would explain why the opiate receptor has evolved?

Writing It Up

With so many questions to be tackled, we did not quite know where to begin. Measuring opiate-receptor binding was such a technically simple procedure that, if we so chose, it seemed that we could attack many of these problems rapidly, without pausing to catch our breath. Instead, we decided to follow another of Axelrod's adages—to tie the project up neatly in an elegant package and publish it. Only then, he said, can you put it behind you and be ready with a fresh, open mind to look at new questions. For Candace and me, the time had come to write up our results.

We felt that the subject was of general scientific interest, but was of widest concern in the United States. The most extensively read scientific journal in America, and perhaps in the world, was the weekly magazine *Science*, published by the American Association for the Advancement of Science, which includes among its members about 130,000 of America's practicing scientists. The majority of the papers in *Science* are brief reports of less than 1,500 words. We would certainly have to be concise. We could include only the most important points from the vast morass of data we had accumulated. Since we were describing an important scientific finding, we had to be particularly careful in choosing our words. While we wanted to convey the general importance of our observations, it was equally crucial to be circumspect and not go overboard in our enthusiasm. Composing this scientific paper was beginning to seem as rigorous a task as executing a perfect sonnet.

In fact, as I recollect, Candace and I put together the first draft in a little less than two hours. We sat at a table with a stack of the key tables and figures before us and discussed what seemed to be the main points. I then dictated into a tape recorder the entire manuscript, pausing only for Candace to interject corrections or her own idiosyncratic phrasings.

This procedure of rapidly dictating a first draft is the one I follow for all scientific papers. My aim is just to get something down. Overly careful composing at an initial stage leads swiftly to writer's block; after only a paragraph or two, weeks of procrastination usually follow. Moreover, I find that obsessively prepared first drafts lack the spontaneity that is so closely linked to innovative thinking. Some senior scientists avoid the first-draft dilemma by turning the task over to the student. Sadly, students often suffer even more than professors from pen paralysis, so that the enterprise may consume months.

After a first draft has been typed up, considerably greater care goes into its editing. The student fills in details, performs statistical calculations, describes the exact experimental protocols so that others can reproduce our experiments, and tabulates a list of literature references. I review and edit this next version, ensuring that the principal concepts are lucidly described. Journals are replete with papers whose jargon-filled prose obfuscates more than it clarifies. I strive valiantly to render scientific writing accessible to as many interested readers as possible.

I found the scientific part of this first paper on opiate receptors easy to write. The results were straightforward and incontrovertible. Clearly we had isolated the site at which the major actions of opiate drugs are exerted. Far more subtle were the challenges in discussing just what our finding implied.

First of all, we were obliged to review how the sites we had identified related to the limited stereospecific binding of radioactive levorphanol that Avram Goldstein had reported the preceding year. Of course, it was subsequently shown that Goldstein had been monitoring interactions of his radioactive drug with well-known tissue lipids rather than the opiate-receptor protein. However, when we wrote our paper, all we knew was that the low 2 percent specific binding Goldstein had observed was quite different from what we had found. In his studies, Goldstein had found the most pronounced differences between the optical isomers of opiates at very high concentrations of the drugs, concentrations a million times higher than the concentration of opiate receptors in the brain. Thus, the drugs were apparently interacting with some tissue constituent which was not a drug or neurotransmitter receptor but which could distinguish optical isomers of opiates. Second, whereas we found marked variations in numbers of opiate receptors in different brain regions, Goldstein had not reported such regional variations. As it later turned out, opiate receptors are highly localized to brain structures which regulate pain perceptions, emotions, and other functions influenced by opiates.

The few sentences in the paper describing Goldstein's research were the most difficult to write. I had great respect for his enormous contributions to pharmacology since the early 1940s and certainly did not want to give offense. Moreover, I worried that Goldstein would feel that he was being upstaged by someone young enough to be his son.

A few weeks after our paper was published I received a letter from Goldstein. "I thought your paper in *Science* was very fine, enjoyed it, and admired your systematic and productive approach to the problem. If I had any criticism, it would be that you were a little ungenerous about the scientific and intellectual precedents of your work. I have found—and being older than you can offer fatherly-type advice—that it enhances rather than diminishes one's scientific stature to lean over backward in acknowledging precedents and priorities. The work you reported would have seemed just as important and significant had you discussed its relationship to our investigations in a more constructive way." Whether justified or not, this criticism stung.

In the section of the paper indicating the general significance of our observations, Candace and I considered a seemingly limitless number of implications. There was no room to describe all of them. Moreover, we felt it desirable to understate matters a bit. I thought back to the classic, very brief report in *Nature* in which James Watson and Francis Crick enunciated the structure of DNA: "We wish to suggest a structure for the salt of deoxyribose nucleic acid (DNA). This structure has novel features which are of considerable biological interest."[1] We decided to follow their lead and end our paper with similarly pregnant phrasing. "Identification of the opiate receptor provides new insight into the mechanism of action of opiates. Our binding assay affords a rapid means of determining the relative potencies of potential narcotic agonists and antagonists, with attendant theoretical and practical implications."[2]

Playing Checkers with the White House

Most scientific manuscripts are copied in triplicate, deposited in the mailbox, and then forgotten, until the editors get back in touch. For the opiate receptor paper, another step was necessary, a political one.

At just about the time we finished our paper, Richard Nixon was on a campaign to impound funds that Congress had appropriated for various activities. The NIH budget was one of several that the White House felt could manage with many fewer dollars than Congress wished to bestow. William Bunney, nominally the director of drug-abuse research at NIH, was being pressured to devote his resources to research that was more politically than scientifically productive. Most medical research is funded by Federal agencies that are reasonably immune to political pressure. But drug-abuse research was being directed straight out of the White House, and this particular administration regarded all things, even the most technical scientific issues, as political fodder. Thus, Bunney had negligible discretion in allocating funds. To spend even small amounts of money, Bunney required permission from the White House Special Action Office of Drug Abuse Prevention (SAODAP). Political scrutiny of research efforts had been intense in 1972 when we fought for creation of the drug-abuse research centers. Now Watergate revelations were appearing almost daily in the *Washington Post*. Potentially disloyal officials were under constant surveillance, and this embraced all the intellectual, liberal types in SAODAP, including Jerry Jaffe, himself a registered Democrat. Hints that any administration official was collaborating with Congressional efforts to augment the NIH budget might be followed by swift, harsh retribution.

It was in this context that I phoned Bunney to tell him the good news about our opiate-receptor findings and to ask if he would like to read the manuscript we had prepared. Was he interested? "My God, yes!" Though a practicing psychiatrist and, at that time, a medical-scientific politician, Bunney has always believed that pure science is the enterprise that matters most. Thus, while Bunney provided excellent advice about how to proceed politically, he also read the paper carefully as a scientist and made several helpful suggestions, including one that caused us to modify our paper in a substantive way. We had been impressed that the relative numbers of opiate receptors in different parts of the brain seemed to match the distribution of the neurotransmitter acetylcholine. In the first version of the manuscript, we strongly suggested that opiate receptors may be found exclusively on acetylcholine neurons. Bunney argued that we were foolish to go out on a limb with such a speculation. Did we not already have enough hard-nosed, concrete data? Why risk conjectures that could turn out to be wrong, in which case critics might focus only on the mistakes and forget about everything we had said that was right? He was absolutely correct. We deleted the theory about acetylcholine. The speculation turned out to be totally mistaken.

At a political level, Bunney suggested that identification of the opiate receptor was just the tonic that support for drug-abuse and general medical research needed at the time. The White House had been arguing that nothing practical ever emerges from basic research, or at least nothing of demonstrated political practicality was likely to emerge during a single presidential term. Bunney saw clearly that the opiate receptor just might be an exception to this rule.

In the first place, if nothing else, many benefits could accrue from a simple and accurate technique for measuring opiate receptors in test tubes. Up to this time, drug companies relied almost exclusively on intact animals to screen chemicals for their potential as psychoactive drugs. To conduct a single experimental test, company scientists would have to employ several doses of the drug in intact rats, using four to six rats at each dose. The skimpiest evaluation of a drug often involved 50 to 100 rats, at $5 a rat. Moreover, 25 grams of the drug had to be synthesized to provide enough to inject all of these rats. A pilot synthesis of a few milligrams, or thousandths of grams, can often be handled in a few days or a week at most; scaling up to amounts a thousand times greater demands a minor feat of chemical engineering that may occupy several weeks or a month of the chemist's time. For the pharmacologist, tests in intact rats are an expensive and time-consuming process.

Contrast all of this work with what would happen if one could screen drug candidates in test tubes with receptor-binding techniques. One need only evaluate the experimental chemical in a few test tubes, ten at most. One milligram of the drug would be more than enough. There would be virtually no animal cost, as the brain of a single rat would suffice to evaluate 5,000 or more experimental drugs. How many drugs could one evaluate in a receptor-binding test? One of our skillful laboratory technicians could process 500 test tubes before lunchtime. In other words, the ability to screen chemicals for potentially active drugs, the first rate-limiting step in drug development, could be speeded up by a factor of 100, if not 1,000.

These practical consequences of receptor-binding techniques were apparent to Bunney and me. They have since been validated in practice. Now, some 16 years later, one can measure receptors for almost all the known neurotransmitters in the brain and for the major psychoactive drugs. Roughly 50 different receptors can be evaluated in simple, efficient test-tube procedures. Drug companies all over the world use these procedures to evaluate potential new drugs.

Because of the Republican party's link to big business, a scientific discovery offering greater efficiencies in developing new products would surely be attractive to Nixon's people. The theoretical possibility of coming to grips in a systematic, scientific way with the ordeal of addiction also should play well in Peoria. And, of course, it could not be forgotten that the opiate-receptor research had been supported by a specific program, the Drug Abuse Center effort, which had emerged directly from the Nixon White House.

Meet the Press

For all of these reasons Bunney felt that "this one should be played big!" He recommended that we call a national press conference, inviting all the media, television, news magazines, and newspapers. We mulled over an appropriate location. Should it be held in Washington? If so, at NIH or the White House? What about having the press conference in Baltimore at Johns Hopkins? In favor of Washington was the fact that a major portion of the national news media is housed there. One obtains more coverage for a Washington news conference than for one held almost any other place in the country. With a Washington news conference, credit would accrue far more directly to the White House than with one held in Baltimore.

On the other hand, Bunney was concerned that the myriad convolutions needed to obtain White House approval of press coverage might complicate matters. He also felt it important to emphasize that this scientific discovery stemmed from university research. Finding the opiate receptor was not a political activity but an intellectual one. Though he wished to stress the beneficial fallout for society, still he did not want to drown the science in an ocean of politics. Accordingly, he concluded that the meeting with the press should take place in Baltimore.

The Johns Hopkins Public Relations Department orchestrated the event with impeccably professional acumen. Bunney came up from Washington to introduce the press conference and place opiate research in a national perspective. He had invited Jerry Jaffe to participate, but Jerry declined, for reasons which to this day remain unclear. Though the senior faculty member usually handles presentation of his laboratory's research to the press, I chose to include Candace. This was a major event in her professional career, and I wanted her to savor the good things that accrue from the many hours of laboratory endeavor.

On March 2, 1973, our paper, "The Opiate Receptor: Demonstration in Nervous Tissue," appeared in *Science*. The press conference was held the same day. Reporters divided their questions roughly equally between scientific issues and political ones, such as the funding of drug-abuse research and how this work meshed with President Nixon's war on heroin.

The overall scope of news coverage was most impressive in

this, my first, encounter with mass-media publicity. Network as well as local TV cameras and reporters were present. *Newsweek* published a feature article on the story. The Associated Press and United Press news syndicates carried reports, and there were features in the *New York Times* and the *Washington Post*. But as clippings from newspapers in the rest of the country trickled into Baltimore, I was struck by the extraordinary capacity of the news media to misinterpret stories. One article in a Memphis paper, presumably derived from wire service reports, was headlined, "Cocaine cure discovered by Hopkins researchers." Of course, we never uttered one single word about cocaine, and the Associated Press and United Press wire dispatches also never mentioned the word. Several other news stories heralded "new heroin cure." I received many phone calls from families of heroin addicts begging to be included in our "treatment program." Of all the things that sadden me about inaccurate press coverage of scientific research, nothing has disturbed me more than these phone calls and letters. Raising false hopes in the minds of seriously disturbed victims and their families is intolerably cruel. It hurt me greatly to have to tell all these people that hoped-for cures were, at best, far distant in the future.

Besides the multitude of letters and calls from drug addicts seeking a cure, and from patients with terminal cancer seeking a nonaddicting painkiller, I received many letters from scientists who had heard news accounts but had not yet read *Science*. Not surprisingly, they were perplexed and sometimes outraged, knowing that we could not possibly have discovered a cure for heroin, or cocaine, or both. It was a real eye-opener to me to realize that distortions and misunderstandings of this type take place in the news media every day; like most scientists, I was naive about how news stories reach the daily papers, and I was accustomed to taking their accuracy more or less for granted. In fact, much is lost in the multiple translations of a story from the background source, to the primary reporter for a wire service, to the local reporter who rewrites the wire-service story, to the editor who modifies the piece to make it maximally attractive to the audience (or to fit the space allotted) and then appends what he or she feels to be the most "gripping" headline. It's all a little like the parlor game of "telephone," where a message is whispered ear to ear in a chain of ten or more half-drunk party mates.

Not all newspeople are monsters, of course. Indeed, some of the most brilliant and conscientious intellects I have encountered emanate from the news sector. Harold Schmeck of the *New York Times* has for decades consistently provided accurate, lucid accounts of medical research. His description of the opiate-receptor work was a model of honest, clear, thoughtful journalism. Matt Clark of *Newsweek* wrote a story so incisive that he presaged our future receptor work more aptly than anything I could have conveyed. He attributed one quote to me which I may have muttered in different words but which was formulated by Clark with such perspicacity that my thinking about future directions was markedly influenced. "'We can assume,' says Snyder, 'that nature did not put opiate receptors in the brain solely to interact with narcotics.'" Reading my own thoughts reassembled so lucidly prompted greater reflection on the possibility of morphine-like substances that might occur naturally in the brain—endogenous painkilling neurotransmitters whose site of action was the opiate receptor.

Another of my "quotes," as set down by Matt, came to me almost as a revelation when I read it: "What we found has given us confidence to look for the receptor sites of other drugs as well."[3] Indeed, a major effort of my laboratory in the ensuing years has been to apply the lessons of opiate-receptor technology to identification of receptors for all the major neurotransmitters in the brain, discovering along the way how many psychoactive drugs act via specific receptors.

Who's on First?

I am no mystic. However, my twenty years in science have nearly brought me to believe in mental telepathy, clairvoyance, psychokinesis, and all the other forms of parapsychology that we scientists officially eschew. As other experienced researchers acknowledge, scientific discoveries have a way of appearing quite independently at the same time in different places. The best-known example of simultaneous discovery in science deals with mathematics' most powerful tool, the calculus. It appears to have been elaborated simultaneously and independently by two of the greatest minds in the history of mathematics, Sir Isaac Newton in England and Baron Gottfried Wilhelm von Leibnitz in Germany. Historians have carefully studied this event, trying to see whether either of these innovators influenced the other. More importantly, they have tried to analyze the happenings in seventeenth-century mathematics that might have set the stage for the two mathematicians' doing their epochal work at the same time.

Similarly, the question of who first authored the theory of evolution by natural selection is hotly disputed. Most authorities agree that Charles Darwin was convinced of a role for evolution by the evidence he collected during his travels aboard the *HMS Beagle* from 1831 to 1836. While reading Malthus' *Essay on the Principle of Population* after returning home from the *Beagle* voyage, Darwin began to formulate the concept of natural selection, but he procrastinated in publishing his theory for some twenty years while he accumulated further evidence. Meanwhile, quite independently, in 1858 the naturalist Alfred Russel Wallace read the same volume by Malthus and developed a concept of evolution that was essentially the same as Darwin's. Wallace then composed an essay on natural selection, which he mailed off to Darwin for comments. Though the *Origin of Species* was not yet complete, the theories of the two naturalists were jointly presented to the Linnean Society on July 1, 1858.

Why did Darwin delay publishing for twenty years, finally putting together his work just as another naturalist developed the identical conception? How was it that two naturalists should quite independently be inspired to think of their key concept by the writings of Malthus, an economist who himself had no interest at all in the diversity of animal and plant species? Historians agree that neither Darwin nor Wallace stole from each other. Clearly, the intellectual climate was ready for evolution, but why such coincidental discoveries?

In modern times, electronic, written, and auditory communication is so rapid that seemingly telepathic events are far more susceptible to rational, unmysterious explanation. Every scientist who makes a discovery stands on the shoulders of those who preceded him. The rapid pace of scientific developments today is such that creative researchers are engaged in almost constant leapfrog; and in fast-moving fields like molecular biology, simultaneous discoveries are more often the rule than the exception. In the long run, it seems that no single person is indispensable to the advance of science, and even the gap that measures how far "ahead of his time" a scientist's work can be said to be continues to narrow. If Newton and Leibnitz had not lived, calculus would surely have been discovered by someone else, though the delay may have been as much as a hundred years; without a Wallace or Darwin, the delay in the conception of evolution by natural selection would probably have been no longer than twenty years. Today, a highly original, genuinely creative researcher working in the field of genetic engineering might at best function intellectually one or two years ahead of his competitors.

Despite many disputes among researchers as to who stole what from whom, it is probably uncommon for outright thievery of ideas to take place in modern research. More frequently, one scientist may be influenced unconsciously by faint glimmers he has obtained from the work of another scientist. This may take the form of a rumor heard from the scientific grapevine or an abstract read casually and set aside. At some later time, without remembering the source, the scientist "independently" divines the crucial experiment, hastens to his laboratory and thence to publication "simultaneously" with the fellow from whom the concept was quite unwittingly pilfered.

Competition at Home and Abroad

Unbeknown to Candace and me, the research of a Swedish pharmacologist, Lars Terenius, had been moving in directions similar to ours. Terenius was well acquainted with the receptor concept, having studied estrogen receptors for several years. He also had been interested in the accumulation of neurotransmitters and drugs, including morphine, by nerve cells in the brain. He had been investigating the possibility that one way in which morphine could be inactivated would be for it to be accumulated into nerve endings by some pump-like mechanism.

Terenius used preparations of nerve-ending membranes from the brain. He obtained these by homogenizing the brain with techniques much like our own to disrupt the cells. Radioactive dihydromorphine, the drug that had been a total failure in our own hands, worked beautifully for Terenius. We now know that our failure stemmed from the rapid destruction of the compound by normal laboratory lighting. In contrast with American labs, which are almost always illuminated by bright fluorescent beams, Swedish laboratories tend to be more dimly lit by incandescent bulbs. Moreover, in Terenius' town of Uppsala, several miles north of Stockholm, during the winter months it is dusk by midday and dark by the middle of the afternoon. Thus, neither bright fluorescent lighting nor the sun shining through the windows would destroy the radioactive dihydromorphine.

To evaluate drug specificity, Terenius compared the optical isomers of the opiate methadone, which is used widely in treating heroin addicts. He was struck by the difference in their behaviors. Levo-methadone blocked the binding of radioactive dihydromorphine much more effectively than dextro-methadone. Knowing that the pharmacologic effects of opiates were stereospecific, Terenius quickly perceived the possibility that what he was monitoring might not be interactions with uptake sites in nerve endings. Instead, perhaps the dihydromorphine was binding to opiate receptors on the outside surface of the nerve ending. He felt that this experimental finding, the inhibition of dihydromorphine binding by the optical isomers of an opiate drug, merited publication. He wrote up his results as a brief report for the Swedish journal *Acta Pharmacologica et Toxicologica*. His paper was received by the journal on November 6, 1972, and published in the spring of 1973.

Thus, working independently, Lars Terenius had observed

stereospecific binding of opiates about the same time that we conducted our work. But did he discover the opiate receptor? One could argue that the findings in his initial paper were too limited to permit the conclusion that he had identified the receptor that mediates the actions of opiates in humans. As Avram Goldstein's experiments showed, commercially available lipids can bind opiates stereospecifically. Gavril Pasternak, a student in our laboratory, subsequently found that even the fiberglass filters used routinely in our research can bind opiates stereospecifically. Because of the nature of his experiments, Terenius was very cautious. He did not announce that he had found the opiate receptor. Instead, he said, "The high-affinity binding material could well be the actual narcotic receptor."[4] In subsequent publications Terenius described the additional experiments necessary to ensure that the binding he had observed involved the opiate receptor. In these later studies he evaluated an extensive range of opiate drugs to show that the more potent ones, pharmacologically and therapeutically, are also more numerous at the binding sites. Nonetheless, even in his preliminary studies, what Terenius had been measuring was, in fact, the opiate receptor.

Terenius was not the only scientist hot on the trail of the opiate receptor at the time of our discovery. Eric Simon, a professor at New York University Medical School, began his career studying the biochemistry of bacteria. After discovering that certain opiates affect protein synthesis in bacteria, he turned his activities more and more toward studies of how opiates act in the brain. Simon attended scientific meetings with pharmacologists more than with molecular biologists. He perceived that much of the ongoing pharmacologic research on opiates was not directed at fundamental questions. He could see that there must exist an opiate receptor and that searching for it should take scientific precedence over other endeavors. He even devised a chemical strategy for purifying opiate receptors if they could ever be identified. Like Candace and me, Simon purchased radioactive dihydromorphine from the New England Nuclear Company and tried to measure its binding to brain membranes. His experiments also were a flop, presumably because of Manhattan sunlight and fluorescent light bulbs. But Simon reasoned, as I had, that perhaps the problem was that the dihydromorphine was not potent enough to fish out the few opiate receptors in the brain from the vast sea of nonspecific binding sites. Perhaps a more potent radioactive opiate would work.

The most potent opiate known to man in 1972, and still today, is a drug called etorphine. In some experiments in intact animals etorphine has proved to be as much as 10,000 times more potent than morphine. It works in such tiny doses that it is actually a hazard for experimentalists as well as for the workers in the factories that manufacture it. Partly because of the hazards of formulating such a potent drug and distributing it widely to hospitals all over the world, etorphine has not been used much clinically. However, it provides the ideal means for immobilizing a large animal, such as an elephant or hippopotamus, for capture and subsequent transport to a zoological park. The hunters merely fire on the beast with darts containing tiny amounts of etorphine.

Simon asked the New England Nuclear Company to prepare a radioactive form of etorphine. He mixed it with homogenates of brain tissue, seeking binding that would be inhibited by opiate drugs in proportion to their pharmacologic activity. As he has recounted to me, Simon detected binding associated with what seemed to be pharmacologically relevant opiate receptors. However, the amount of binding was small and not readily manipulated experimentally. Perhaps the relative proportion of tissue, fluid, and chemicals was not optimal. Simon, like Goldstein before him, was trapping the radioactive etorphine bound to membranes by centrifuging the mixture so that the brain membranes containing opiate receptors with bound radioactive drug would spin down into a pellet, which was then monitored for its radioactive drug content. With such a procedure the membranes are not washed to remove nonspecifically bound opiates. This technique contrasts with the filtration techniques we used, which permit rapid but thorough washing.

Upon reading our paper on the opiate receptor in March

1973, Simon immediately adopted our binding techniques, except that he used etorphine instead of naloxone. His experiments then worked like a dream. In less than a month Simon obtained highly reproducible, accurate opiate-receptor binding. He was able to show that the ability of a substantial number of drugs to compete for etorphine binding paralleled their known opiate-like activities in man. He rapidly completed a scientific paper. It was received by the *Proceedings of the National Academy of Sciences USA* on April 19, 1973, and published in the July 1973 issue.

On the Boardwalk

During the intoxicating month following publication of the Pert–Snyder paper in *Science*, I was totally unaware of Simon's activity. That April, the Federation of American Societies for Experimental Biology was to have its annual meeting in Atlantic City. The FASEB meetings are the major yearly get-together of basic biomedical researchers. The societies comprising FASEB include those representing American physiologists, biochemists, pharmacologists, pathologists, nutritionists, and immunologists. In the early 1970s the FASEB meetings typically attracted 15,000–20,000 young and old researchers, from Nobel laureates to first-year graduate students. Audiences at the major symposia number in the thousands.

In early March the program of the FASEB meeting was mailed out. I noticed that a fairly large symposium on receptors was scheduled. It included lectures on several hormone receptors, the acetylcholine neurotransmitter receptor, and a talk by Eric Simon on "Strategies for Identifying Opiate Receptors."

I assumed that Simon planned to speak on the conceptual approaches one might, in theory, employ to investigate opiate receptors. Such a speech seemed superfluous now that we had already identified these receptors. Surely it would be in the best interests of scientific discourse for our research to be included in the program. But FASEB symposia are tightly scheduled, with a specific number of minutes, usually 15 or 20, for each speaker and five minutes for discussion. Might it be feasible to squeeze in an extra, brief talk?

I did not know the symposium chairman, Edward Reich, a professor at Rockefeller University, and so I was reluctant to make any overtures. However, my boss, Paul Talalay, chairman of the Johns Hopkins Pharmacology Department, was a longtime close friend of Reich's. He phoned. I was allotted five minutes and warned not to exceed the allocation even by thirty seconds *or else*.

I was sitting in the audience somewhat agitated at the prospect of pouring forth at machine-gun pace an hour presentation in exactly five minutes. I was not paying much heed to the initial symposium speakers. Then Simon stepped to the podium and, predictably, proceeded to review the evidence about the action of opiate drugs that would prompt a reasonable person to propose the existence of opiate receptors. He described briefly some of our work, as presented in our *Science* paper, which had appeared just one month before. I came abruptly to attention when he next proceeded to present his own experiments with etorphine.

Later, on the boardwalk, Simon handed me an advance copy of his manuscript. The description of his experiments indicated that he had used the method of "Pert and Snyder." I assumed from this that he had initiated his studies subsequent to reading our paper; and in an article I wrote shortly thereafter I described our findings and the independent observations of Lars Terenius and then stated that Simon and collaborators had "subsequently confirmed" these findings.

Simon was angered, and some months of ill will ensued. While the experiments reported in his paper used our techniques, Simon explained to me, he had, in fact, made the initial observations some months earlier with his own procedures.

The dispute was long ago, and Simon and I are now friends. But just such subtleties of interpretation and misunderstandings are the stuff out of which scientific controversy is often born.

· 5 ·

Agonists and Antagonists

The publicity in the lay press and the controversy in the scientific community that followed the initial reports on the opiate receptor prompted innumerable invitations to lecture at universities and at scientific meetings. It was tempting to become a jet-set scientist. But having been blessed with two daughters, then 3 and 6 years old, and a wife, Elaine, who has never been bashful about expressing her views, I was not in much of a position to become an absentee father and husband, even if I had wished to. Besides, I have seen many scientists seduced by a seemingly glamorous lifestyle, only to find later that, far from advancing their career, the scurrying about ultimately destroyed all that had been productive in their lives. The scientist who leaves his laboratory bench for long is not likely to continue creative work. While senior scientists may hope that students can handle experiments on a day-to-day basis with supervision every couple of weeks, this arrangement seldom works out. Too much time away breaks a scientist's emotional ties to the drama of discovery.

So many avenues were now suddenly open to exploration, it was difficult to chart a course. One of our first moves was to find out whether the general strategy that made it possible to measure opiate receptors might be applicable to receptors for other neurotransmitters of the brain.

Just a few months after the first experiments labeling opiate receptors, Anne Young, an MD-PhD student in our lab, succeeded in measuring receptors for the neurotransmitter glycine. Glycine modulates motor neurons in the spinal cord and thus may mediate the spasticity of patients suffering strokes. Soon after Anne's findings, Hank Yamamura was able to measure receptors for acetylcholine. Acetylcholine neurons are selectively lost from the cerebral cortex of individuals with Alzheimer's disease. Measuring receptors for the neurotransmitter dopamine has provided clues to the action of drugs that relieve symptoms of schizophrenia (see Chapter 9).

Of course these new research areas did not distract us from the many questions still to be answered about opiate receptors. To investigate whether opiate receptors develop before birth, Candace Pert purchased pregnant rats and evaluated the brains of the fetuses. Sure enough, even during fetal development opiate receptors are detectable. Clearly, one can become addicted to opiates in the womb.

Ontogeny recapitulates phylogeny—that is, during the development of the embryo, an organism repeats the same stages that occurred during its evolution. This fact led us to wonder how early in evolution opiate receptors appeared. This straightforward question was assigned to a bright high school student, David Aposhian, who spent a summer examining all sorts of animals—mice, guinea pigs, fish, even lobsters, cockroaches, and fruit flies—using our techniques. David found opiate receptors ubiquitously distributed among vertebrates. Even the hag fish, the most primitive vertebrate available to us, had as many opiate receptors as humans. Opiate receptors were clearly not an attribute linked to high intellect. Invertebrates such as insects, on the other hand, did not seem to have opiate receptors. We now know that even invertebrates possess opiate receptors but of a unique subtype that could not have been detected with the radioactively labeled naloxone we were employing in those days.

All these developments were invigorating, but I was distressed that we were not making inroads into the fundamental question of what differentiates agonists from antagonists. Both bind to the same receptor, but whereas the agonist triggers a change in cellular function, the antagonist does nothing at all. It merely sits on the receptor, blocking access of agonists. Agonist and antagonist opiates are chemically similar, yet vastly different in their pharmacologic effects. Why?

When a neurotransmitter or a drug agonist binds to a receptor, it alters cellular function by changing the neuronal membrane's permeability to one or another ion. How recognition of the neurotransmitter changes the movement of an ion through the neuronal membrane is a basic question of neurobiology. Communication between neurons via neurotransmitters is the event unifying all information processing in the brain. If we could understand how this event is mediated, we would have proceeded well along the path to understanding how the brain "works."

Drugs have proved to be powerful probes for exploring the neurotransmission process. If we could find some molecular difference in how agonists and antagonists interact at receptor sites, that difference might be a clue to how recognition at receptors changes ion movement. If we could solve this problem for the opiate receptor, the answer might apply to all neurotransmitter systems.

I was perplexed when our first experiments showed no differences between the way agonists and antagonists bind to opiate receptors. The agonist morphine and the antagonist nalorphine, a derivative of morphine, bind to the opiate receptor with virtually identical potency. Candace and I looked very carefully at every nuance of how agonists and antagonists influenced the opiate receptor, but we could find no differences.

Mixed Blessings

Besides theoretical issues about brain function, there were practical reasons for attacking this question. Some of them are best understood in light of the history of how opiate antagonists were developed. In 1915 the German pharmacologist J. Pohl synthesized a very simple variation of the codeine molecule. The nitrogen of codeine and other opiates is a critical part of the chemical, as it forms a major site for attachment to the opiate receptor. Codeine possesses a methyl group, a carbon with associated hydrogens, attached to this nitrogen. Dr. Pohl synthesized a molecule in which the methyl group of codeine was replaced with an allyl group, consisting of three carbons instead of the single carbon of a methyl group. This allyl derivative of codeine blocked the ability of morphine or heroin to depress the breathing of animals. A striking feature of Pohl's finding is how such a tiny change in the structure of a drug would transform agonist into antagonist.

Pohl never sought a practical application for his discovery. The rest of the medical and scientific community also took little note. Twenty-five years elapsed before the pharmacologist Klaus Unna examined a similar drug. Like codeine, morphine also possesses a nitrogen with an attached methyl group. When Unna replaced the methyl group with an allyl group (see figure 5), just as Pohl had done with codeine, the drug blocked the pharmacologic effects of morphine. Nalorphine, as this drug was later called, proved to be substantially

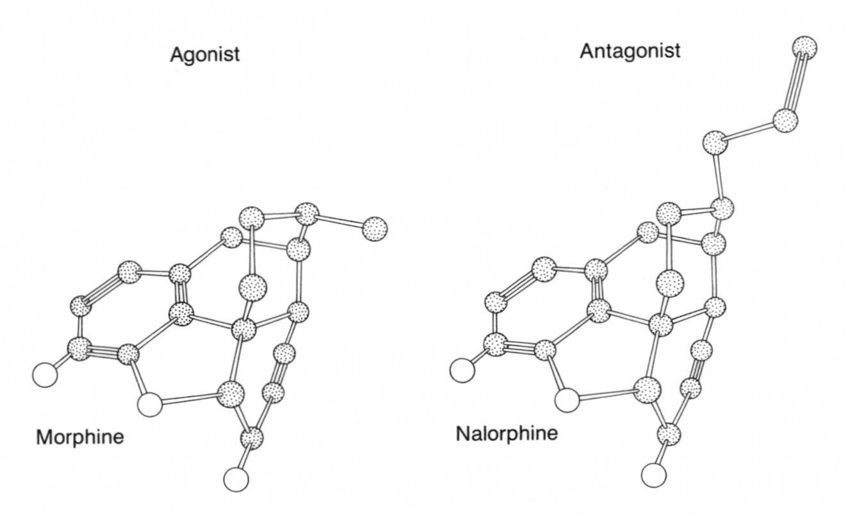

Figure 5. The opiate agonist morphine is quite similar in its chemical structure to the antagonist nalorphine.

more potent as an opiate antagonist than the allyl codeine derivative. Although one would have expected therapeutic applications to follow shortly, it was not until 1951 that nalorphine was used as an antidote for morphine poisoning in humans. Abraham Wikler, an eminent narcotic researcher at the Federal Narcotic Hospital in Lexington, Kentucky, then gave nalorphine to former heroin addicts. First he addicted his subjects to opiates such as heroin, morphine, or methadone. (Since the patients were Federal prisoners, they were pleased to participate in these experiments, even if they involved drug administration, partly in the hope that cooperation would bring them more favorable treatment.) Then Wikler injected nalorphine and immediately witnessed severe opiate withdrawal symptoms. In other words, Wikler showed that nalorphine could displace morphine from opiate receptors. Wikler also found that as doses of nalorphine were increased, nonaddicted subjects developed considerable anxiety and a feeling of severe mental discomfort which often progressed into a psychotic state. Why the blocking of opiate receptors leads to psychosis was not at all apparent to Dr. Wikler and, indeed, is still a mystery.

The most important chapter in this story came in 1954, when two of the fathers of clinical pharmacology, Drs. Louis Lasagna and Henry Beecher, working at the Massachusetts General Hospital in Boston, began to explore in fastidious detail how opiate agonist and antagonist drugs might interact in patients. Their experimental model was relatively simple. They chose patients with pain following various surgical procedures, and treated them with various doses of morphine by itself or in combination with different doses of nalorphine. Just as predicted, nalorphine blocked the effects of morphine, in proportion to the doses of the two drugs employed. As a "control," they treated some patients with nalorphine by itself, expecting it to do nothing at all. To their surprise, nalorphine, when administered alone, also relieved the patients' pain—in some instances just as effectively as morphine. How could a drug which everyone agrees to be an opiate antagonist succeed in easing severe postoperative pain? One must conclude that nalorphine, an opiate antagonist, also possesses some agonist activity.

Enter the Drug Companies

The therapeutic implications of this unexpected finding did not elude the major drug manufacturers. They knew already that opiate antagonists such as nalorphine are not addicting. If some of these antagonists also have agonist action, they might represent the long-sought-for nonaddicting painkillers. Lasagna evaluated a wide range of nalorphine doses in patients with pain. Though he could consistently observe analgesia with the drug, he found, as Wikler had previously noted, that with doses high enough to relieve severe pain, patients became apprehensive and anxious, and they sometimes developed delusions and hallucinations. Clearly, nalorphine was not going to solve the problem of addiction-free pain relief.

Next, numerous drug companies took up the search for a "mixed agonist-antagonist" opiate. Nobody knew just why nalorphine caused its mental side effects; perhaps one could come up with an analogous drug that would be free of side effects. The key might be just the right proportion of agonist and antagonist. This proportion might not be quite right in the nalorphine molecule.

But how would one go about identifying such drugs? Screening for antagonist activity in intact animals is relatively simple, if crude. One gives a dose of the experimental drug to an animal and then immediately gives a dose of morphine. If the experimental drug is truly an antagonist, it will block the ability of morphine to relieve pain. Measuring pain relief in rats or mice involves subjecting them to a painful stimulus, usually associated with heat, and then measuring the time that elapses before the animal acts to avoid it. For example, how long will a rat stay on a hot plate before jumping off? Or how many seconds go by before a mouse or rat flicks its tail to avoid the heat of a hot light?

In screening tests of this type, the antagonist actions of

nalorphine were readily apparent—that is, animals injected with nalorphine followed by morphine still jumped off a hot plate or flicked their tails very quickly. But the analgesic effects of nalorphine were not apparent in these crude animal models of pain.

Drug companies eventually developed screening tests in animals that could detect mixed agonist-antagonist opiates. In one model, acetic acid, the active component of vinegar, is injected into a rat's abdomen to elicit pain. Mixed agonist-antagonist opiates such as nalorphine reduce this type of pain to some extent, but the test is relatively nonspecific; many nonopiate drugs, including aspirin, are also effective.

Using the acetic acid test, the Sterling-Winthrop Drug Company did come up with a mixed agonist-antagonist drug that would relieve pain in humans at doses that usually do not cause dysphoria and psychosis. Pentazocine, marketed under the trade name of Talwin, was the first clinically relevant mixed agonist-antagonist opiate painkiller. Because of its antagonist properties, pentazocine is less addicting than morphine. Moreover, at higher doses it is less likely than morphine to depress respiration.

At the time of the opiate-receptor discovery, pentazocine was the only mixed agonist-antagonist opiate on the market. Because of its lesser addictive liability, pentazocine was quite popular and was one of the best selling of all opiates. However, it had its drawbacks. Pentazocine was not as effective as morphine in relieving extremely severe pain. And though at therapeutic doses it did not cause dysphoria and psychosis in most patients, a significant minority of patients developed these unpleasant side effects. Accordingly, there was still a great impetus in 1973 for the pharmaceutical industry to come up with an improved mixed agonist-antagonist. But experimental systems using intact animals to identify this drug continued to pose a problem. The acetic acid test was far from ideal. Also, pharmacologists were unable to determine from experiments in intact animals just how the agonist and antagonist components of a drug's activity interacted in producing therapeutic doses and side effects. They had no means of quantifying the exact proportions of agonist and antagonist actions in a test drug. It was clear that understanding agonist and antagonist activity at a molecular level would have considerable therapeutic potential.

Aversion Therapy

In 1973 there was also a market for new drugs that would be "pure" antagonists. A pure opiate antagonist is simply one that occupies the opiate receptor, blocks all the actions of conventional opiates, and yet causes no pain relief or euphoria. In 1973 the only pure opiate antagonist marketed was naloxone, still the major agent used in treating opiate overdose. Because it is virtually inactive when taken by mouth, naloxone is administered almost exclusively by injection, usually intravenously. Such a route of administration is fine, indeed preferable, for treating opiate overdose. When a drug is injected intravenously, its onset of action is almost instantaneous. Another advantage of naloxone is that its effects last for only a couple of hours in doses conventionally employed to treat opiate overdose; most physicians are leary about inducing prolonged withdrawal symptoms in a patient whose body may be loaded with massive amounts of heroin.

But in 1973 there was one potentially important clinical use for a pure opiate antagonist for which the short-acting effects of intravenous naloxone were not appropriate, and that use was in Richard Nixon's war on heroin. The cornerstone of Jerry Jaffe's national plan for combating heroin addiction in the United States was a nationwide network of methadone-maintenance clinics. However, many people objected that these clinics were merely substituting addiction to heroin with addiction to methadone. True, methadone is a long-acting opiate agonist that can be taken by mouth, whereas heroin is short-acting and is used only by intravenous injection. Still, keeping victims addicted to opiates for several years or more hardly seemed like a definitive solution to the opiate-addiction problem. One of Jaffe's longer-range goals was to devise a means of treating heroin addiction that would not require the use of opiate agonists.

He and other workers in the opiate field felt that pure opiate antagonists might just be the answer. Orally effective, long-acting pure opiate antagonists could perhaps break the vicious cycle of heroin addiction by blocking the euphoria that comes within seconds of plunging a needle into one's arm. If a person were treated with a long-acting pure opiate antagonist, even massive doses of heroin injected intravenously would fail to produce a high. After a period of time, so the theory went, the addict would lose interest in heroin. Such a strategy is analogous to the use of Antabuse (disulfiram) in the treatment of alcoholism. Antabuse interferes with enzymes that metabolize ethanol. If an alcoholic takes Antabuse regularly and drinks whiskey, ethanol metabolism will be altered so that the chemical acetaldehyde will accumulate in the blood. Acetaldehyde causes a severe pulsating headache, nausea, vomiting, chest pain, blurred vision, and confusion. Presumably this aversive conditioning causes the alcoholic to lose interest in drinking. However, Antabuse works only if the alcoholic continues to take it. If a drinker has enough motivation to take Antabuse regularly, then he or she probably is well enough motivated to participate in an Alcoholics Anonymous program and become totally abstinent without any drug. Likewise, the only heroin addicts who would ingest the long-acting pure opiate antagonist regularly might be those with sufficient motivation to quit heroin even without the antagonist.

Regardless of these concerns, development of an orally effective, long-acting, pure opiate antagonist was a major priority at the White House. Jaffe called top executives of the large drug companies together for frequent meetings to plot strategy, and he promised Federal money to subsidize their research efforts; the drug companies, after all, could expect little profit from a drug used primarily for heroin addicts, who are not noted for paying medical bills. Though the White House office on drug-abuse prevention was chary in its support of basic research, it made available almost limitless funds for research on pure opiate antagonists. Indeed, much of the money for the drug-abuse research centers was carved out of the enormous budget that had been allocated by Congress for a crash program on narcotic antagonists.

Dividing the Spoils

At the prospect of so many rewards, both in basic science and in drug treatment, we had ample motivation to work hard studying the molecular basis of agonist and antagonist actions at opiate receptors. For several months we tried everything imaginable, but with no success. A crack in what seemed an impenetrable barrier came about serendipitously.

So many opiate-receptor-related projects beckoned that Candace could not handle the load by herself. I assigned a new technician in the laboratory, Adele Snowman, to work with Candace full time. Fortunately for us, Adele turned out to be the most remarkable technician I have ever known. Candace or any other hard-working student in the laboratory could run through about 200 test tubes in the course of a day; Adele could easily handle 500, and sometimes even a thousand. Not only was she productive in sheer volume of experimental work, she was also more technically skillful than anyone else. The precision of her values was uncanny. Duplicate samples that Adele would measure agreed far more closely than similar pairs measured by any of the PhD students. With such accuracy it was possible for us to detect small differences between experimental groups. Besides her technical virtuosity, Adele displayed a devotion to work rarely seen even in PhD candidates. Adele would routinely appear in the laboratory at 5:30 a.m., and, if the experiment demanded, she would work at night. I would often go to the laboratory on Sundays and discover Adele working furiously at the lab bench "to get a few things out of the way."

The teaming of Candace and Adele enabled us to deal with many questions in a brief period of time. Still, I felt that there were so many problems to be addressed that another scientist at the PhD level should be involved in the opiate-receptor work. Gavril Pasternak, a medical student who was also working toward a PhD, joined the laboratory in late 1972. Gavril entered my office for our initial discussions just about the time I was deciding that Candace could not handle all the opiate projects by herself. I asked Gavril if he would like to do his PhD on the opiate receptor. He was delighted at a virtual guarantee of a successful PhD thesis, one which would surely result in numerous publications.

Candace and Gavril differed greatly in their approaches to science. Candace repeatedly manifested flashes of intuitive brilliance; by contrast, Gavril worked methodically and cautiously. He repeated experiments three or four times and analyzed the results of each experiment over and over again, hoping to dissect out significance from each small item. He rarely resorted to the elaborate theorizing that was Candace's forte.

I thought it might be a good idea for two scientists with completely different approaches to focus on the opiate receptor. In this way each could make discoveries based on his or her unique strength. On the other hand, I did not want the two of them to work together as a team on identical projects, since this would steal from each the sense of personal, unique creativity. A discovery feels good only insofar as the scientist senses deep down, "This is mine, mine, mine!"

How was I to divide up the opiate receptor? Candace was already exploring how opiates exert their clinical effects, questions which I would call "pharmacologic." There remained many questions about the biochemical nature of the opiate receptor. Was it a protein? Were lipids involved in its functions? Could the opiate receptor be separated from brain membranes and purified, much as enzymes are, and then isolated as a single molecule? It seemed to me that Gavril would be ideal for these biochemical investigations. As an undergraduate, he had already worked on biochemical projects and so had considerable technical expertise. Additionally, such detailed biochemical explorations would benefit greatly from someone who was careful, precise, and, most of all, patient.

At first Candace was a little miffed that part of the opiate-receptor kingdom would be removed from her purview. However, I pointed out that her name was already identified with the major discoveries. Moreover, the projects she was addressing promised the biggest breakthroughs. It seemed unlikely that anything dramatic would emerge from analyzing the biochemical features of the opiate receptor. And finally, Candace had at her disposal the services of an extraordinary technician as well as ample funds and equipment for whatever research she wished to undertake. She was in an enviable position for a graduate student, commanding resources that are usually available only to full-time faculty members.

Candace agreed to the arrangement, and the two graduate students proceeded on independent studies of the opiate receptor.

The Sodium Effect

At first this partition worked out reasonably well. However, after about a month, friction between Gavril and Candace commenced and gradually escalated. Besides carrying out his own work, Gavril repeated all the basic experiments on the opiate receptor that Candace had previously undertaken. His argument was that replicating her findings was necessary so that he would be sure he was using the same standard system. Candace felt that Gavril was checking out everything she had done, searching for mistakes that might discredit her in my eyes. Fairly soon I found myself devoting too much time to mediating disagreements between the two students.

Despite the distress attendant upon the Gavril–Candace interaction, there were benefits. The intensely critical attitude that each brought to bear on the experimental work of the other sharpened all of our thinking about crucial questions. And one conflict between Gavril and Candace contributed to a most important breakthrough.

Gavril was evaluating the influence of a wide variety of biochemical reagents upon opiate-receptor binding. He wanted to identify the chemical nature of the opiate receptor, whether it be protein, lipid, or carbohydrate. One reagent that affects some proteins is ethylenediamine-tetraacetic acid (EDTA), which interacts with metals on some enzymes. When Gavril added EDTA to his incubation mixtures, radioactive naloxone's binding to the opiate receptor doubled. When Adele, representing Candace's team, tried out EDTA, there was no effect. Since EDTA did not seem to me to be a terribly important reagent, I didn't greatly care *what* it did to the opiate receptor. However, for Candace and Gavril, EDTA provided another opportunity for confrontation.

As happens so often in science, upon careful scrutiny this apparent conflict in experimental results disappeared. One glaring difference between Gavril's and Adele's experiment accounted for the discrepancy. Gavril had used an opiate antagonist (naloxone) as the binding agent, whereas Adele had used a pure opiate agonist (dihydromorphine, which is closely related to morphine in structure). The conflicting results were both valid: EDTA decreased the binding of the agonist to opiate receptors but increased the binding of the antagonist.

As we began to ponder why EDTA would influence differentially the binding of the two radioactive drugs to the opiate receptor, another headache emerged. As Adele tried to repeat the experiment on successive days, she had difficulty obtaining the same results. On some occasions EDTA did not affect opiate-receptor binding at all. Further detective work by Gavril revealed that the disparity in results on different occasions stemmed from the use of different bottles of EDTA. Like most inorganic chemicals, EDTA is manufactured as a salt formulated together with a positive metal ion, either sodium or potassium. On the days when EDTA lowered binding of the agonist and raised binding of the antagonist, Adele had been using the sodium salt of EDTA. Her "failures" had come on the days when the potassium salt was used instead.

Things suddenly began to make sense. First, we finally had a way to differentiate between opiate agonists and antagonists. Second, the differentiation was not due to EDTA but to the sodium ion. Third, the effect of sodium was selective, since potassium was ineffective. The implications were tremendous. It seemed that sodium might be the ion that "translated" recognition of opiates at the receptor into changes in cellular function. Our results suggested that in the nervous system agonists normally do something to sodium at the receptor, while antagonists either have the opposite action or no action at all.

When we realized that the differentiation by EDTA of agonists and antagonists at the opiate receptor had been due to sodium ion, a number of earlier puzzles were easily resolved. In our first studies of the opiate receptor we had evaluated numerous ions and reported that sodium had very little effect upon opiate-receptor binding, which in those experiments was measured with an antagonist. When Eric Simon published his paper, he reported that sodium ion reduced opiate-receptor binding of the agonist etorphine. We had not paid much attention to the discrepancy in the reports of the two laboratories. In his paper Simon had pointed out the difference between his results with sodium ion and ours, but he assumed that the difference in sodium's effects on opiate-receptor binding of antagonists versus agonists was due to general ionic strength and not to sodium itself. Simon suggested in his publication that "should the latter difference be a general one between agonists and antagonists, it may reflect the difference in receptor binding that results in their distinct pharmacological properties." Simon apparently did not conduct experiments to follow up this suggestion, so he seems not to have taken the discrepancy between his results and ours any more seriously than did we initially.

Did the "sodium" effect really reveal a universal difference between all opiate agonists and all opiate antagonists, or was it limited to the two drugs naloxone and dihydromorphine? We obtained radiolabeled versions of several opiate antagonists and agonists. Sodium reproducibly lowered the binding of all these agonists and either failed to affect or increased the binding of all the antagonists. Of course, by this time we were not even including EDTA in the incubations, as it was clear that the effect was restricted to sodium and, indeed, occurred at very low concentrations of sodium, such as are found in the body. Moreover, it was highly specific for sodium, not being apparent with potassium or other ions.

Having confirmed that sodium ion did selectively differentiate agonists from antagonists, we formulated a simple model. We proposed that sodium ions are crucially involved in communicating the influence of opiate agonists upon cells that possess opiate receptors and that this influence has to do with the passage of sodium through specific ion channels. Evidence that has accumulated in the succeeding ten years suggests that sodium affects the cell in a different way, though it is still possible that specific channels for sodium are involved. In any event, the general principle continued to hold up. We were able to extend this finding to receptors for other drugs, in which specific ions, often sodium, would differentiate between the binding of agonists and antagonists. These influences have provided valuable tools for researchers to explore the question of how neurotransmitter recognition changes cellular function, and just why agonists are agonists and antagonists are antagonists.

Identification of the sodium effect had other payoffs. We hoped that it would provide a means for differentiating between pure agonists, pure antagonists, and mixed agonist-antagonists. One way of doing this would be to obtain radiolabeled forms of all test compounds and determine whether their binding to the opiate receptor was influenced by sodium to a pronounced extent, to a lesser extent, or not at all. Unfortunately, obtaining each experimental drug in a radiolabeled form would pose insuperable logistical problems, not the least of which was the immense expense.

The solution to this dilemma was proposed by Candace one afternoon. She suggested that we use only radioactively labeled naloxone. Its binding would not be affected by sodium. We could then test the ability of nonradioactive experimental drugs to compete with radioactive naloxone for binding sites. A pure agonist should become a very much weaker competitor in the presence of sodium, while a pure antagonist should remain potent. One would expect mixed agonist-antagonists to behave in an intermediate fashion. Within a week Candace and Adele carried out the proposed experiments, with great success. Pure antagonists such as naloxone did not become weaker at all relative to radioactive naloxone when one added sodium to the incubation medium. On the other hand, pure agonists such as morphine became up to fifty times weaker, relative to radioactive naloxone, in the presence of sodium. Agonists with a limited amount of antagonist activity, such as nalorphine, became about one-half as potent in the presence of sodium. The fascinating group of drugs, the mixed agonist-antagonists, especially those that already had shown clinical promise as less-addicting analgesics, became three to seven times weaker in the presence of sodium.

Clearly, we now had a simple test-tube method to identify opiate drugs along an agonist-antagonist continuum. With less than a milligram of an experimental drug, one could evaluate its "sodium index" in an hour using a few test tubes. This contrasted with the week or more of tests in intact animals of different species, tests which required the chemist to synthesize tens of thousands of milligrams of drug at far greater expense and to subject many animals to needless suffering. Screening potential opiates at the opiate receptor in the presence and absence of sodium soon became a standard procedure in the pharmaceutical industry.

Pain and Euphoria

The regulation of opiate receptors by sodium taught us a great deal about what goes on at the molecular level at synapses in the brain. However, it did not shed much light on some of the big questions about how opiates act. We still did not know why opiates relieve pain, or why they cause euphoria, or even how they constrict the pupils of the eye. Why are opiates so potent in depressing respiration, so much so that doses only a few times greater than those needed to relieve pain kill by terminating the breathing process?

Of course, these kinds of questions are fundamental ones for the great majority of drugs that influence brain function.

Why does marijuana make people high? Why does LSD cause psychedelic experiences? The answer that pharmacology professors have offered their students over the years is that presumably the drugs attach to target sites in the brain that regulate those processes influenced by the drugs. This response is almost as tautological as a politician's evasion. The important question is just where are those sites in the brain? Throughout the history of pharmacology, there has been no definitive way to rigorously identify sites of drug action. One could make lesions in different areas of the brain and ascertain if the drug's effects changed. But brain lesions rarely influence only a single group of neurons. Instead, they also affect nonneuronal cells as well as long neurons that happen to be passing through the site of the lesion enroute elsewhere.

It occurred to us that measuring drug receptors might offer a solution to the problem of where the sites of action are. One would merely determine the exact location of receptors within the brain, on the assumption that the most enriched sites carry out functions affected by the drugs.

In our very first experiments we had dissected the rat brain into different areas and found marked variations in levels of opiate-receptor binding. The rat has a small brain, so that we could not dissect discrete areas with known functions. Our findings were tantalizing, but an approach with much greater discrimination would be required to link the location of receptors to function.

After we had detected the regional variations in opiate-receptor binding within the rat brain, Candace and I were eager to examine a species in which one could separate discrete, small regions. Human or monkey brain would be ideal. Neither Candace nor I knew very much neuroanatomy, and we certainly did not know how to dissect a human or monkey brain into the small zones that would be required for a meaningful linking of location to function. For help, we turned to Dr. Michael Kuhar, a new faculty member with expertise in neuroanatomy.

In a brief period Kuhar dissected the brains of five monkeys into forty-eight discrete areas whose functions were reasonably well understood. He similarly dissected two human brains obtained from the Baltimore City Morgue. Candace and Adele measured opiate-receptor binding in all of them and found differences from one to another that were even more interesting than what we had seen in the rat.

Parts of the limbic system were *loaded* with opiate receptors. The limbic system comprises a group of structures that were known to mediate emotional behavior. One portion of the limbic system, the amygdala (see figure 6), had far and away the highest number of opiate receptors in the brain. It would be reasonable to suppose that receptors within the limbic system were associated with the effects of opiate drugs upon mood.

For many years scientists had been exploring areas in the brain that are involved in the perception of pain. Candace and I found that some of these were also greatly enriched in opiate receptors. For instance, the periaqueductal grey zone in the midbrain, a key area in pain perception, had one of the higher levels of opiate receptors. Electrical stimulation in the periaqueductal grey relieves pain. Indeed, several neurosurgeons have used electrical stimulation in this area to bring about fairly long-lasting pain relief in patients with intractable pain from terminal cancer.

The thalamus serves as a conduit for all sorts of sensory information enroute to the cerebral cortex. The lateral thalamus contains fibers that convey the sensations of touch and pressure, whereas the medial thalamus is enriched in fibers specific for pain perception. We found that the medial thalamus possesses one of the highest densities of opiate receptors in the brain, more than triple the number in the lateral thalamus. Thus, it would be reasonable to suppose that opiate receptors in the medial thalamus and the periaqueductal grey have something to do with relief of pain by opiate drugs.

Visualizing Opiate Receptors

At this point it was clear that Candace had done more than enough work to justify the PhD degree. Her thesis defense

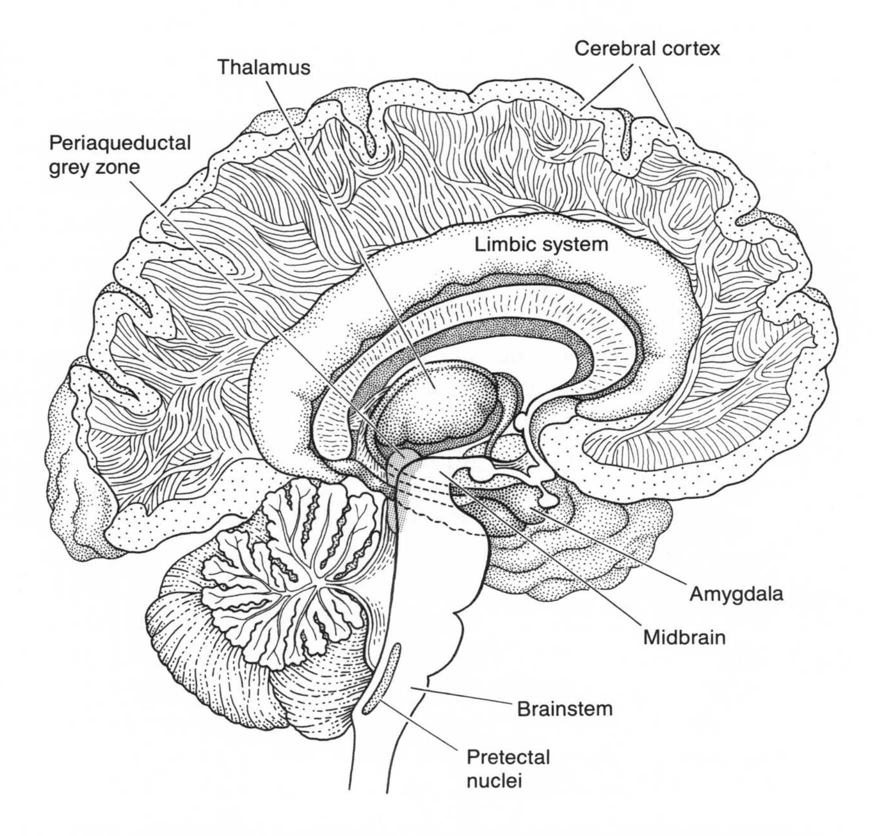

Figure 6. Drugs that affect the central nervous system, including opiates, have their effects in selected areas of the brain. The painkilling action of opiates occurs when the drug binds to neurons in the periaqueductal grey zone and in parts of the thalamus. Opiates produce euphoria when they bind to neurons in the limbic system of the brain, particularly the amygdala.

went smoothly and noncontroversially. Faculty who had expressed concern two years earlier now congratulated her. Everyone acknowledged that she was extraordinarily talented. It was time for Candace to decide what to do for her postdoctoral training.

In years long past, PhD students would seek a faculty position as soon as they received their doctorate. With the coming of ultraspecialization in research, the PhD degree itself no longer procures a good university job, at least not in the biomedical sciences. Most PhDs go on to postdoctoral training, generally in laboratories other than those in which they obtained their doctorate. Candace had to juggle a number of factors. Her husband was obligated to remain in the Army for another year or two, so that Candace could not leave the Baltimore area. Moreover, work on the opiate receptor was so exciting that it would seem a shame to leave Baltimore without following up the numerous leads she had generated during the course of her doctoral work.

In agonizing over just what to do, a proposal emerged that would satisfy Candace's complex needs. Why not remain at Hopkins, working on opiate-receptor problems in my laboratory, but in collaboration with Michael Kuhar, learning new anatomical approaches to the brain and the opiate receptor? The new approach would involve literally visualizing opiate receptors in the brain at a microscopic level. The technique to be employed is called autoradiography.

Pierre and Marie Curie were among the first people to notice that radioactive molecules—in the form of a nugget of uranium in a desk drawer—caused silver grains to develop on adjacent x-ray film. Since their discovery, autoradiography has been used extensively to investigate all sorts of molecules. Why not opiate receptors in the brain? With autoradiography we could determine the location of receptors at a microscopic level. This would provide us with the capacity to identify extremely small nuclei of the brain with a sensitivity and precision thousands of times greater than in dissections of monkey and human brain.

Autoradiographic studies of a radioactive drug bound to a receptor in the brain presented some major difficulties. For instance, one must ensure that the radioactive drug in the brain is associated with the opiate receptor. Drugs elicit their pharmacological effects by binding to specific receptors, but this does not mean that 100 percent of the drug in the brain is bound to the receptor. If only 20 or 30 percent of the drug is bound to the receptor, then autoradiographic studies would visualize, in large part, drug molecules that were not on the receptor.

Candace's idea of working in collaboration with Kuhar to visualize opiate receptors came at a particularly propitious time. He had just completed his first study demonstrating the feasibility of visualizing a neurotransmitter receptor in the brain by autoradiography. Together with Hank Yamamura, Kuhar had been able to visualize acetylcholine receptors in the brain. He used a radioactive drug which, when administered to the rat, bound so tightly to the receptors that essentially all the radioactive drug in the brain was associated with the acetylcholine receptors.

Despite this success, it was clear to all of us that the opiate receptor would represent an even greater challenge. Candace and I had just completed a study trying to label opiate receptors in intact rats with radioactive naloxone injected intravenously. Despite the fact that naloxone binds tightly to the opiate receptor, only about 20 or 30 percent of the radioactivity detected came from molecules bound to the opiate receptor. Clearly one would be unable to carry out autoradiography with radioactive naloxone. Searching the published literature for drugs that might be more potent than naloxone, we identified diprenorphine, which then was, and still is, the most potent opiate antagonist on earth. We went to considerable effort and expense to obtain diprenorphine in a radiolabeled form. When Kuhar injected it into a rat, virtually all the radioactivity in the brain was bound to opiate receptors. We were in business!

Under Kuhar's tutelage, Candace then mastered the very tedious procedures involved in slicing frozen sections of the brain and preparing them on microscope slides. We waited almost two months for silver grains to develop on the photographic emulsion. The pictures were well worth the wait. Opiate receptors were concentrated in specific zones, confirming what we had found in our earlier studies. There were dense clusters of silver grains representing opiate receptors in the periaqueductal grey, but now we could see that they form a discrete circle elegantly circumscribing the boundaries of the periaqueductal grey.

Other new findings helped clarify the relation of opiates and pain. Sensory information is conveyed by nerves that pass from the limbs and body surface to the dorsal, or backside, part of the spinal cord, where the sensory nerves terminate. In this zone, referred to as the substantia gelatinosa because it is gelatin-like in appearance, there exist many very short neurons. These receive information entering from the sensory nerves and pass it back and forth to each other. The substantia gelatinosa is the first way station where genuine integration of sensory information, including information about pain, takes place within the central nervous system. We witnessed dense bands of opiate receptors confined to the substantia gelatinosa of the spinal cord.

The discovery of opiate receptors in the spinal cord ended a debate that had divided pharmacologists for many years. Some believed that all of the painkilling effects of opiates were exerted in the brain. Others had obtained experimental evidence that opiates could relieve pain at the spinal-cord level. Our findings showed that both groups were right. The opiate receptors in the substantia gelatinosa would explain pain relief at the spinal-cord level, while receptors within the brain would account for "supraspinal" analgesia.

In numerous areas of the limbic system we found high densities of opiate receptors, consistent with our earlier observations in dissections of monkey brain. The autoradiographic studies enabled us to pinpoint much more precisely the exact structures involved and account more meaningfully for the euphoria that opiates produce.

We were even able to explain why opiates constrict the pupils of the eye. Certain nuclei in the brainstem, such as the pretectal nuclei, regulate pupillary diameter. These were invariable "hot spots" for opiate receptors. The ability of opiates to depress the breathing process could also be better understood. Various nuclei of the brainstem related to the vagus cranial nerve control breathing reflexes. Most of the nuclei of the vagus nerve were enriched in opiate receptors. One of the most striking of these was the nucleus of the solitary tract, prominent in regulating blood pressure as well as breathing.

Visualizing the precise location of opiate receptors in discrete areas of the brain convinced even the greatest skeptics that these receptors serve important functions. People began to wonder why they were there.

· 6 ·

The Brain's Own Morphine

As early as the summer of 1973 Candace and I were puzzling over why opiate receptors exist. We were intrigued by the ability of sodium to differentiate between opiate agonists and antagonists. Sodium must somehow be involved in how opiates change neuronal functioning in the brain. Neurotransmitters excite or inhibit neuronal firing by opening or closing pores in the membrane of the neuron specific for various ions. Presumably sodium was an ion whose passage through the cell membrane was crucial for opiate effects. In this way the opiate receptor was behaving much like the receptor for a neurotransmitter.

The opiate receptor resembled neurotransmitter receptors in other ways. Amounts of opiate receptors varied considerably in different brain regions. Similarly, most neurotransmitters are not uniformly distributed throughout the brain. Instead, the pathways that contain transmitters tend to be highly localized to specific brain regions. Some neurotransmitters are contained in only one or two pathways. Other neurotransmitters occur in dozens of pathways. However, in almost all instances the concentration of a neurotransmitter varies markedly throughout the brain. Naturally, one would expect numbers of receptor sites for neurotransmitters to vary in parallel with the numbers of neurotransmitter neurons themselves.

The striking regional variations in numbers of opiate receptors fit excellently with everything we knew about neurotransmitter receptors. Could the opiate receptor normally serve as a binding site for some hitherto unknown neurotransmitter? We doubted that animals had evolved opiate receptors exclusively to deal with properties of the poppy plant.

There were other sound reasons for suspecting that the brain makes its own morphine-like neurotransmitter. A psychologist, David V. Reynolds, in 1969 found that electric stimulation in the brainstem of rats could cause pain relief. John Liebeskind, professor of psychology at UCLA, and his students David Mayer and Huda Akil explored in greater detail Reynolds' initial observation and came up with a number of startling findings. They stimulated the brains of rats in different areas and found that the best analgesia was obtained when they stimulated the periaqueductal grey, a part of the brainstem which, we later discovered, is loaded with opiate receptors. Liebeskind showed that the analgesic effect could not be explained by nonspecific influences on motor activity or overall responsiveness of the rats. His most extraordinary observation was that the opiate antagonist naloxone blocked the stimulation-induced pain relief. It seemed that pain relief came about because electrical stimulation caused certain neurons to fire and release an opiate-like neurotransmitter which in turn acted upon opiate receptors that could be blocked by naloxone.

Though some of the papers on pain relief through electrical stimulation had been published in 1971 and 1972, biochemically oriented researchers such as I were unaware of this work, even in late 1973. Yet it provided some of the most compelling evidence for the existence of a morphine-like neurotransmitter. Had we known about this work, we doubtless would have begun to search for the opiate receptor and its neurotransmitter years before we did. Like so many scientists, we suffered the punishments as well as the rewards of specialization.

False Starts

Candace and I discussed at length how to seek such a hypothetical morphine-like factor, the brain's own morphine. How would we measure it? Our one special tool was the opiate-receptor binding technique. Anything that was morphine-like in the brain ought to compete with radiolabeled opiates for binding to the opiate receptor. In a typical experiment one may find 1,000 counts of radioactive naloxone bound to opiate receptors in brain membranes. Adding a certain amount of a morphine-like factor obtained from brain extracts might lower this to 600 counts. Thus, one would have a simple system for monitoring the amount of morphine-like factor. But how were we to prepare such brain extracts? One certainly would not want to add a completely crude homogenate of the brain. Reasoning that a morphine-like factor might have chemical properties similar to morphine itself, we wondered if such a factor might be extracted into an organic solvent such as ether or butyl alcohol. Accordingly, Candace mixed ground-up brain together with organic solvents and conducted experiments to see whether material that had passed into the organic solvents would inhibit the binding of radioactive naloxone to opiate receptors. To our excitement, she found activity.

The excitement was short-lived. Showing that an extract inhibits the binding of naloxone to opiate receptors does not prove that one is dealing with a new neurotransmitter. Extracts of brain and almost any animal tissue contain millions of chemicals, some of which will tie up naloxone in one or another chemical bond so that it is not available for the opiate receptor. Such a chemical would deceive one into believing that it was a morphine-like factor. Of course, the chemical would not have been interacting with the opiate receptor at all but only with the radioactively labeled drug. Other chemicals might actually interact with the opiate receptor, but in a nonspecific fashion. Any assay based on *inhibition* is therefore inherently risky. After all, one can inhibit naloxone binding to the opiate receptor by dropping the test tube to the ground, spitting in it, or adding any noxious chemical.

How were we to ascertain whether the inhibition of naloxone binding that Candace observed reflected a biologically meaningful morphine-like neurotransmitter? One trick would be to take advantage of the fact that the numbers of opiate receptors differ greatly in various brain regions. Presumably the amount of the morphine-like factor should show similar regional variations. Additionally, a morphine-like neurotransmitter should occur in the brain but not in peripheral tissues such as the liver that have no opiate receptors. Candace investigated whether there were differences in the extent of inhibition of naloxone binding with extracts from different areas of the brain and liver. But the initial forays failed. Liver extracts did just as well as, if not better than, brain extracts in competing for naloxone binding. Moreover, the amount of the material inhibiting binding was the same in all the areas of the brain we examined.

However, we did not conclude that a morphine-like neurotransmitter did not exist. We had only tried one way of extracting the brain. Perhaps the brain's "morphine" has a chemical nature that is different from the morphine of poppy plants—one that would not dissolve in organic solvents. One could look in water extracts of the brain. A true opioid neurotransmitter might have been masked by a large amount of nonspecific material that interacted with naloxone or the opiate receptor. We could try fractionating the brain into many distinct, relatively pure chemical components to see if one of them influenced opiate-receptor binding in a selective fashion. A variety of other strategies were also possible. The question we had to ask ourselves was whether we wished to invest the energy required to pursue the project right then.

Myriads of important and exciting questions about the nature and function of opiate receptors remained to be explored. All of these were relatively straightforward, readily attacked, and practically guaranteed to give results. Should we drop all of this and turn our energies to the search for a new substance that might not even exist? I told Candace to drop the endogenous morphine problem and to work instead on safer projects.

Surprise from Scotland

Every field of science has its cliques, small groups of researchers with similar interests who keep in close touch with each other, organize scientific meetings together, and congregate with one another more often than they see their neighbors down the block. Such fraternities provide much-needed emotional support for researchers working in areas that are too arcane to be easily explained over the back fence. Opiate researchers are one such community.

Having never worked in the opiate field until Candace and my first experiments with the opiate receptor in late 1972, I was an outsider to this group. One might have thought that our work identifying and characterizing opiate receptors would be welcomed by narcotic researchers. To the contrary, some were openly antagonistic. They were so skeptical of biochemical work that many refused to believe we had measured sites on cell surfaces that account for the therapeutic effects of opiate drugs. Even those who accepted the discovery of the opiate receptor still maintained that opiates must do much more than just binding. A frequent refrain went much like this: "One doesn't really know what a drug is doing until the response is measured in an intact animal. The only way to study opiate receptors properly is to measure pain responses in rats, or, better yet, monkeys and humans."

Difficulty in viewing the world in a new way is sometimes thought to be a by-product of advancing age. At least one member of the opiate fraternity was a decided exception to this stereotype. Hans Kosterlitz turned 70 in 1973. Along with many other Jewish scientists, he had left Germany on rather short notice in 1933, bound for the United Kingdom. Many of his colleagues obtained academic refuge in Oxford or Cambridge, the only two universities in Great Britain where state-of-the-art biomedical research was being conducted at the time. But Kosterlitz did not travel to these academic meccas. He had developed a strong interest in diabetes, a productive area of medical research at that time. Frederick Banting, Charles Best, and John MacLeod had received the Nobel Prize in Medicine in 1923 for their extraction of insulin from the pancreas and their demonstration that insulin could reverse the otherwise fatal course of diabetes. In 1933 John MacLeod was chairman of the Physiology Department at the University of Aberdeen. Located far in the north of Scotland, Aberdeen was virtually inaccessible in 1933, physically as well as intellectually. But if one wanted to learn about diabetes, sugar metabolism, and insulin, what could be better than associating with Professor MacLeod, even in Godforsaken Aberdeen?

Six months after Hans and Hanna Kosterlitz arrived in Aberdeen, John MacLeod died. This being the height of the Depression all over Europe as well as the United States, academicians had little job mobility. Scientists, like most people, were thankful for any job whatsoever. Scientifically, Kosterlitz was as marooned as Robinson Crusoe. Nonetheless, he managed to make a series of valuable contributions to our understanding of sugar metabolism. Gradually, in the 1950s, Kosterlitz turned his attention to the effects of drugs on smooth muscle of the intestine and became interested in how morphine influences intestinal contractions to elicit its constipating effects.

Upon his retirement from chairmanship of the Pharmacology Department in 1968, Kosterlitz was able to devote himself full time to research. He developed techniques for identifying opiates as pure agonists, pure antagonists, or mixed agonist-antagonists. Drug companies became interested in his simple smooth-muscle techniques, and soon Kosterlitz was an important fixture of the international opiate research community and a consultant to several drug companies.

Until Candace and I commenced our work on the opiate receptor, I had never heard of Hans Kosterlitz. As I began to read the literature in the field, I became aware of his considerable contributions. However, as late as 1973 I still had not met him.

In November of 1973 Fred Worden, the Executive Director of the Neurosciences Research Program, asked me whether I felt a work session on the opiate receptor was a reasonable idea. The NRP comprises a group of some 30–40 associates from many different disciplines, some totally unrelated to the brain, who share an interest in the nervous system. It is an elite group, with up to half a dozen of the associates at any one time being Nobel laureates. The NRP regularly sponsors workshops of 15–20 scientists to discuss specific areas in depth. I was very enthusiastic about the possibility of bringing together researchers in the opiate field. Even if we did not come up with brilliant syntheses or ideas for innovative new directions, I would at least have a chance to meet the major opiate scientists. Worden asked me to identify four or five workers in the field to attend a planning session in December, with a view to holding the formal workshop in the spring of 1974. On my list of key opiate researchers was Hans Kosterlitz.

I invited Kosterlitz to visit Baltimore and give a seminar at Johns Hopkins the day before our planning session in Boston. The logistics worked out magnificently. Hans and Hanna had hoped to visit the United States anyhow and spend Christmas with their only son, Michael, a theoretical physicist then at Cornell University in Ithaca. My wife, Elaine, and I quickly became good friends of the Kosterlitzes. There was no generation gap. If anything, Hans and Hanna were more vigorous than Elaine and I. Despite the fact that Aberdeen time is five hours ahead of Baltimore time, on the night they arrived the Kosterlitzes insisted on staying up in spirited conversation, mixed with ample doses of Hans's gift of Glenmorangie Scotch malt whiskey. We got to bed at 2:00 a.m.

In my laboratory and at the planning session in Boston Kosterlitz argued strongly that one of the fifteen participants in the work session should be a young assistant-professor colleague of his named John Hughes. He was a little mysterious about the reasons for including Hughes, at first saying only that "John has some findings which I think you will find to be most interesting." He finished this statement with a slight twinkle in his eye and an audible chuckle.

I cannot abide secrets. I implored him to stop behaving like a tease, whereupon he confided that Hughes and he had evidence that they were able to measure a morphine-like substance that occurs naturally in the brain. I told him of the frustrating misadventures in our laboratory when we attempted to do the same thing. He assured me that in Aberdeen all the appropriate control experiments had been carried out. What they were measuring was indeed the real McCoy.

Brookline, May 1974

In 1973–74 the Neuroscience Research Program was housed in the Brandegee Mansion in Brookline, a suburb of Boston. The Brandegee family made its fortune in the whaling business in the nineteenth century. Their house, with its massive gardens, is one of the great mansions of America, competitive with the most opulent stately homes of England. Twice a year the associates of the NRP gathered in this setting to debate subtleties of brain activity. The workshops, on the other hand, dealt with less ethereal experimental advances in specific areas of neuroscience.

On May 19, 1974, some 30 scientists met for a workshop on "Opiate Receptor Mechanisms" (see figure 7). Naturally, there was much spirited discussion about the opiate receptor, but the historically significant event of the three-day meeting did not relate to the opiate receptor itself.

John Hughes revealed what he and Kosterlitz had been up to for the past several months. At that time their laboratory had no expertise in using radioactive drugs to measure opiate receptors. However, they were quite adept in employing smooth muscle to monitor the actions of opiates. Accordingly, they asked the question, "Do brain extracts possess a substance that mimics the effects of morphine upon smooth muscle?" Morphine inhibits the electrically induced contractions of several smooth-muscle systems. One of the most extensively studied is the intestine of the guinea pig. Hughes and Kosterlitz had also developed the vas deferens of male mice as a model of opiate action. The vas deferens is the tube that conveys semen from the testes to the penis. Contractions of the vas deferens account for ejaculation, which is triggered by the firing of nerves to the vas deferens. These nerves use as their neurotransmitter norepinephrine, the neurotransmitter of sympathetic nerves throughout the body that are activated

Figure 7. Members of the Neuroscience Research Program, May 19–21, 1974. *Seated (left to right)*: Gavril Pasternak, William Bunney, John Hughes, Hans Kosterlitz, Steven Matthysse, Francis O. Schmitt, Solomon H. Snyder, Avram Goldstein, E. Leong Way, Vincent P. Dole, and Horace Loh. *Middle row (left to right)*: L. Everett Johnson, Frederic G. Worden, Robert D. Hall, Candace D. Pert, Yvonne M. Homsy, Parvati Dev, Huda Akil, Floyd E. Bloom, Agu Pert, Peter A. Mansky, William H. Sweet, Albert Herz, William R. Martin, and Harriet Schwenk. *Top row (left to right)*: Ian Creese, David J. Mayer, Eric J. Simon, Leslie Iversen, Diana Schneider, Pedro Cuatrecasas, A. E. Takemori, Arnold J. Mandell, Arthur E. Jacobson, Jose M. Musacchio, and Lars Terenius.

during excitement and stress. Norepinephrine released from these nerves speeds the beating of the heart, causes perspiration, increases muscular strength, and dilates the bronchial tree to facilitate deeper breathing during stressful physical activity. Presumably an excessive release of norepinephrine from the sympathetic nerves to the vas deferens accounts for premature ejaculation in nervous young males who are overly excited or overly anxious when confronting a naked female.

One can readily stimulate the norepinephrine-containing nerves of the vas deferens with a simple electrical stimulator. Morphine inhibits the electrically induced contractions of the vas deferens by acting at opiate receptors on the nerves to impair the release of norepinephrine. The relative potencies of numerous opiates in inhibiting these contractions correlates closely with their potencies as painkillers. Moreover, opiate actions on the vas deferens are blocked by low concentrations of the antagonist naloxone. Thus, the opiate receptor of the vas deferens appears to be closely similar, if not identical, to that of the brain.

In their first experiments, Hughes and Kosterlitz had no difficulty showing that brain extracts would inhibit electrically induced contractions of the vas deferens. But like Candace and me, they were worried about the possibility that the effects of the brain extracts might be nonspecific. All sorts of general trauma and noxious chemicals will inhibit contractions of smooth muscle. To test whether they were dealing with a biologically meaningful opioid neurotransmitter, the Scottish researchers added naloxone to the tissue bath in concentrations that would have blocked the effects of morphine. Nothing changed; the brain extracts continued to inhibit the contractions. Clearly the inhibitory effects of the brain extract were nonspecific.

But unlike Candace and me, Hughes and Kosterlitz did not quit at that point. They purified the brain extracts. With relatively simple and modest purification steps they obtained material whose blockade of electrically induced contractions *was* reversed by naloxone.

This was it. Hughes and Kosterlitz felt that they had identified the long sought for opioid substance of the brain. They worked furiously to purify the material. By the time of the NRP meeting, fairly extensive purification had been carried out. Moreover, they had some evidence that the concentration of the morphine-like material varied in different areas of the brain in general agreement with regional variations in numbers of opiate receptors. They even had some hints as to the chemical nature of the substance, at least in general terms. Certain enzymes that cleave peptides rapidly destroyed the morphine-like activity. Thus, it seemed apparent that the material they were studying was a peptide. This explained why our own attempts to extract the material with organic solvents had been a failure. Peptides are chains of amino acids linked together as in proteins, which are just very large peptides. Because of their electric charges, peptides simply do not dissolve in organic solvents. Hughes and Kosterlitz even obtained a rough estimate of the molecular weight, suggesting that the peptide contained five to ten amino acids.

The Race Is On

After Kosterlitz and I had discussed the presumed endogenous opiate back in December 1973, my interest in doing something about it in our own laboratory was again piqued. Since Candace was heavily involved in other experiments, I felt that this might be a suitable project for Gavril. He seemed to have the patience that would probably be required to isolate from the brain a chemical that we expected would comprise less than a millionth by weight of brain tissue. Gavril liked the challenge, and we had several discussions on appropriate strategies. He conducted a number of exploratory experiments. After the Brookline meeting in May 1974, Gavril moved into high gear.

Gavril and I were not the only ones to join the race with Kosterlitz and Hughes. In Uppsala, Lars Terenius had begun looking for an endogenous morphine-like factor. Since Terenius had considerable experience with opiate-receptor binding techniques, he approached the problem very much as we did. First, he attempted to see whether the brain contained a substance with a chemical structure resembling morphine that could be extracted into organic solvents. Like Candace and me, Terenius found nothing biologically meaningful using this approach. He then made water extracts of the brain and detected material that competed for opiate-receptor binding. He even conducted some initial purification steps and was able to estimate the molecular weight. He felt that the material was probably a peptide of about the same size as what Hughes and Kosterlitz had suggested.

Terenius described his results at the meeting in Brookline. I do not know exactly when he had begun his experimental work. However, it is possible that he had initiated his studies as early as had the Scottish workers and certainly before we had obtained positive results. But Terenius did not go further and attempt to isolate and obtain the chemical structure of the morphine-like factor.

Like Candace, Gavril searched for a substance in the brain that would compete for opiate-receptor binding. However, he knew better than to try to extract the material with organic solvents. We reviewed different ways of extracting the brain to seek a water-soluble substance, presumably a peptide. In my ten years of experience in extracting water-soluble compounds from the brain, I had considerable success with simply homogenizing the brain in water and boiling it to precipitate protein and other insoluble "junk." Such a water extract can be crystal clear and free of many of the impurities that bedevil biochemical measurements. In boiled brain extracts Gavril did find a considerable amount of material that competed with radioactive naloxone at opiate receptors. Then came the crucial control experiments. Would all parts of the brain have the same ability to compete for opiate-receptor binding? To our great delight, we found very pronounced differences in the amount of morphine-like factor in different brain regions. The differences were so pleasing that I could hardly believe them and demanded that Gavril repeat the experiments several times. The relative concentrations in different brain regions of the morphine-like factor, which we abbreviated MLF, very closely paralleled relative numbers of opiate receptors.

There were no a priori reasons for assuming that the brain's endogenous opiate would be an agonist or an antagonist. We had open minds about this rather important issue. It made a fine subject for cocktail-party conversation among biologist-types. Would one expect the poppy plant to *mimic* a substance that occurs naturally in vertebrates, in effect enhancing the painkilling and euphoric effects of the endogenous neurotransmitter, or would it *block* effects that the endogenous substance might otherwise have? Of course, the experiments of Hughes and Kosterlitz showing that brain extracts mimic morphine's effects already provided evidence that the substance was an opiate agonist.

Gavril conducted a series of experiments which confirmed that we were dealing with a substance that had opiate agonist activity. Low concentrations of sodium markedly weakened the ability of MLF to compete for naloxone binding to opiate receptors. Potassium had no effect at all. This fit elegantly with our earlier findings that the same low concentrations of sodium inhibit interactions of opiate agonists with the opiate receptor, while potassium is inactive. Thus, agreeing with Hughes and Kosterlitz, we concluded that the endogenous opiate was an agonist.

In other ways Gavril's results also agreed with those of the Scottish workers. He was able to estimate the molecular weight of the substance and came up with values essentially the same as did Hughes. The sensitivity of MLF to various enzymes also indicated that we were dealing with a small peptide containing, we thought, five to eight amino acids.

But since the Scottish researchers could already measure the morphine-like factor and were trying to purify it, why did we bother to enter the race at all? Why should we join in the search for the presumed endogenous opiate, when Hughes and Kosterlitz had already conducted the initial experiments demonstrating that it existed in the brain? Wasn't one laboratory working on the problem enough? These questions go right to the heart of research strategy. For some scientific problems the answer might be yes, one lab is enough. However, with an extremely important research quest, science is better served when several laboratories attack the question with different tools. The initial identification of an endogenous morphine-like material was only the tip of the iceberg. The next important tasks were to isolate the material, determine its chemical structure, and characterize its function. This series of goals could keep several laboratories hard at work for a number of years.

When several laboratories attack the same problem, resulting competitions may be bitter or friendly, depending on many factors, especially the people involved. For ourselves and the Scots, friendship always prevailed.

· 7 ·

The Name Game

There were no permanent people in Julie Axelrod's laboratory. Most of the laboratory chiefs at the National Institutes of Health devoted major portions of their time to the seemingly endless bureaucratic contortions necessary to obtain permanent slots for their employees. Life is so much easier with people who never leave and thus represent pillars in the lab, able to carry on the traditions of years past and to show new people the ropes. Axelrod eschewed all these traditions. After he won the Nobel Prize in 1970, he could have had as many permanent scientists in his group as he wished. He wished to have none. Instead, he preferred temporary students, postdoctoral fellows who would stay for only two or three years, then leave for faculty positions in universities and be replaced by a new handful of young scientists. Although changing personnel every two to three years meant a great deal of extra work for the lab chief, Axelrod was willing to shoulder these extra burdens in order to see new young faces and new ideas. Every student in the laboratory had his or her own personal research project. The frequent turnover of students would force Axelrod to reevaluate every year or two just where science was moving or ought to be moving and where he should turn his own efforts.

I have always run my own laboratory in the same way. There are no permanent doctorate-level researchers. My lab is populated by eight to ten postdoctoral fellows, PhD students, and MD-PhD students. Gavril Pasternak was in this last group. Because he had not been in our lab at the beginning, he did not share in the discovery of the opiate receptor. Though overshadowed in this way by Candace, Gavril was nevertheless extraordinarily productive. During his time in the laboratory Gavril characterized many of the biochemical properties of the opiate receptor. He showed how a variety of treatments can distinguish between the ways that agonists and antagonists interact with the receptor, differences which are still the focus of research in labs all over the world.

Most important of all, Gavril devised a strategy for identifying the morphine-like factor in the brain. He characterized its properties extensively enough to establish unequivocally that this was the morphine-like neurotransmitter, the normally occurring substance in the body that interacts with the opiate receptor. Using the simple and sensitive technique he developed for measuring this material, Gavril monitored its distribution in many different regions of the monkey's brain, which further established its biological relevance.

After three years in the laboratory, the time came for Gavril to leave. He had selected his specialty field, neurology, and had arranged for an internship and neurology residency at Johns Hopkins. On July 1, 1975, he departed.

With most incoming students I spend much time reviewing potential new projects. I learn what the student has worked on in past years and where he would like to direct his efforts. I describe what we have been doing in a variety of areas. We then go back and forth, toying with potential ideas for new work and settle on whatever seems most promising, regardless of whether the idea originated with myself or the student. For Rabi Simantov things were clear-cut from the outset.

Rabi obtained his PhD at the Weizmann Institute of Science, the pinnacle of scientific institutions in Israel. Israel is replete with extremely bright young Jewish boys and girls interested in science, but Israel is seriously limited in financial resources to support their work. Only the most talented and hard-working students can secure admission to the Weizmann Institute and emerge with a PhD degree. Among these, Rabi was a star. During the three years of his doctoral research, which dealt with molecular mechanisms in cancer, Rabi turned out more publications than many university professors.

Rabi arrived in our laboratory just a few months before Gavril departed. Among his skills in many areas, Rabi was proficient in purifying proteins and peptides. Gavril had done some preliminary purification of the morphine-like factor. Thus, there was little question as to just what Rabi should assume as his major research effort for the two years he planned to be in our lab. Though Rabi had never done any brain research in the past, he began at a breathtaking pace and never slowed down for twenty-four months. It was immediately apparent that Rabi had a "green thumb," able to make all sorts of technically complex experimental maneuvers succeed with what seemed like no effort at all. His breadth of knowledge in different areas of biologic research was extraordinary. Rabi had the ability to dredge up from his broad scientific background all sorts of disparate facts and come up with creative syntheses. Needless to say, he was also hardworking, routinely putting in twelve- and fourteen-hour days.

Candace had just completed her PhD work and was spending about half her time across the hall in Mike Kuhar's laboratory, working on the autoradiographic techniques to visualize the opiate receptor. Supertechnician Adele Snowman was now assigned to work with Rabi. They were an amazing team. I cannot recall any other period when so much solid work was done in so short a time.

Cornering the Market on Calf Brains

Knowing that the body contains a substance with a particular activity advances our understanding, but only to a limited extent. For many years scientists knew that various glands possessed hormonal activity. The real breakthroughs came when the hormones were isolated in pure form and their exact chemical structures were determined. Then it was possible to develop therapeutic agents by synthesizing new chemical structures with appropriate modifications of the natural hormone. Only by knowing the chemical nature of the natural substance could one study the enzymes that form and degrade it and thus seek new drugs that act by blocking one of these enzymes. Without establishing the chemical structure of a substance, one cannot develop sensitive and specific techniques for measuring its levels in body tissues and fluids and determining abnormalities in disease states.

Attempting to isolate a morphine-like entity from the brain was certainly not the first effort to isolate biologically active peptides. A classic example of peptide isolation from the brain had to do with the so-called hormone-releasing factors. Their isolation had required the brains of 250,000 pigs or calves. When Gavril showed that he could measure reliably the brain's morphine-like factor, I recalled the precedent of the hormone-releasing factors. Perhaps we would require enormous supplies of brain tissue. How were we to go about cornering the market on large animal brains?

Just about this time I had begun visiting the Sandoz Drug Company in Basel, Switzerland, as a consultant. Dr. Stephan Guttmann, then director of chemical research, had become interested in using receptor-binding technology to screen new drugs. In the course of my discussions with the Sandoz people, I asked whether they might be able to assist in obtaining large quantities of calf brain and carrying out the initial processing, concentrating the extracts to manageable volumes that could then be mailed to Baltimore. They were most accommodating, and Dr. Daniel Hauser at Sandoz commenced a search for optimal sources. It turned out to be a complex logistical operation. Finally, in late November 1975 Dan succeeded in lining up most of the calf slaughterhouses in Denmark and devising a means to extract the brains and send them to us.

Everybody agreed that isolating and working out the structure of the morphine-like peptide was of paramount importance. We knew that Hughes and Kosterlitz had already made much progress in this effort. Terenius seemed to have dropped out. Clearly, we would have to work swiftly to catch up with our Scottish competitors. Rabi did not dawdle, waiting for Danish brains. He obtained a few calf brains from a Baltimore slaughterhouse and commenced to work. As things turned out, Rabi needed only 20 or 30 brains for all his activities. We never used the Sandoz gift.

Peptides are made up of amino acids linked to one another. Proteins are simply very large peptides containing one hundred or more amino acids in sequence. The specificity of information conveyed by a peptide or protein comes from the exact sequence of the various amino acids. The body contains twenty amino acids. These can be arranged in almost any sequence within peptides or proteins. The range of information that can be conveyed by peptides and proteins can be calculated with simple high school mathematics. For instance, in a five amino-acid peptide there are several million possible amino-acid sequences. It is not surprising that nature chose this method of chemical structure to convey all the complex, specific information inherent in life on earth. First we would have to isolate an absolutely pure opioid peptide. Only then could we try to work out its amino-acid sequence.

Rabi's general strategy was to separate brain extracts into many different fractions and then monitor the ability of each of the fractions to compete for radioactive naloxone binding to opiate receptors. If all the receptor-competing activity was localized to a few fractions, we would thereby have succeeded in separating the active components from a large number of inactive brain chemicals. We would then try a different procedure to further purify the active fractions.

Rabi's purification of the morphine-like material moved quite rapidly. Using conventional techniques for purifying peptides, Rabi could separate the brain extract into a large number of fractions, each containing varying amounts of peptides and other chemicals in the brain. Each of the fractions was tested for its ability to compete with the binding of naloxone to opiate receptors. Within several weeks Rabi had purified the opiate-like material about 400-fold.

At this point I began worrying whether what we were studying was really the true morphine-like factor of the brain. Our earlier confidence had been based on Gavril's finding that the distribution of the substance throughout the brain paralleled the distribution of opiate receptors. However, there are a number of neurotransmitters with fairly similar distributions. We had not yet shown that our material had morphine-like pharmacologic activity. The approach that Hughes and Kosterlitz were employing was somewhat more reassuring, as they were testing whether the material would mimic morphine's effects on a genuine, almost-alive piece of muscle. Best of all, they could check each response for specificity by determining whether naloxone would block the effects of the brain extract.

It occurred to me that we might be able to test the biological activity of our partially purified morphine-like factor even more rigorously than Hughes and Kosterlitz. We might be in a position to test whether our substance would relieve pain. All it took was a call to Agu Pert, Candace's husband.

Agu's Rats

While Candace was working with me, Agu was fulfilling his military obligation in the Army at the Edgewood Arsenal, a research facility some twenty miles outside of Baltimore where the Army has conducted much of its "pharmacologic" warfare studies. Agu was given a large lab and sufficient facilities to do almost anything he wished. Of course, he and Candace had discussed many times her research with the localization of opiate receptors, and Agu was curious about sites in the brain where morphine exerts its various effects. Injecting tiny amounts of morphine into discrete areas of a live monkey's brain, he discovered that pain relief was obtained most effectively and with the smallest doses of morphine when the injection was placed into the periaqueductal grey. Of course, the periaqueductal grey turned out to be greatly enriched in opiate receptors.

In the summer of 1975, when Rabi was carrying out his purification work, Candace finished her training at Hopkins. She and Agu moved to Bethesda, Maryland, where they both obtained positions at the National Institutes of Health. They were able to continue their work on opiate mechanisms and, in fact, were stationed in the same laboratory group. I telephoned Agu and asked if he might be able to test our extracts for their analgesic effects in rats. Agu was eager to collaborate.

Rabi gave Agu samples containing various fractions he had obtained from the brain. Some of the fractions contained a high density of morphine-like material as measured in our receptor-binding technique. Other fractions had no opiate-like activity, even though they had substantial amounts of total peptide material. We took care to ensure that Agu was "blind"—that he did not know which fractions had been positive for opiate-like activity.

I remember well witnessing the first successful experiments. Agu had piled cages of rats in his car and had driven them 50 miles from NIH to Hopkins. Each rat had a small tube implanted in his skull so that a drug or other substance could be injected directly into the periaqueductal grey of the rat's brain. When Agu injected morphine through the tube there ensued an impressive sequence of events. First, the rats would leap as much as five feet in the air, squeal, and rotate in a peculiar way. This intensely hyperactive period was followed by a prolonged period of sedation during which the rats were clearly unresponsive to painful stimuli. We didn't even need a complex experimental apparatus to detect the analgesia. One could pinch the rat's tail intensely, ordinarily an extremely painful stimulus, and observe no response at all.

When Agu injected fractions of the brain extract containing material that was 400-fold enriched in the morphine-like factor, the animals behaved just as if they had been injected with morphine. We witnessed the same leaping and squealing, followed by sedation and a lack of response to painful stimuli. This moment was clearly one of the "highs" of that remarkable year's work.

We conducted all sorts of control experiments to prove that what we had observed was real. When Agu administered the brain fractions that did not compete for naloxone binding, the rats showed no behavioral response at all. When he pretreated the rats with the opiate antagonist naloxone, they no longer responded either to morphine or to the opiate-like brain extract. When Agu moved the tube in the rat's skull so that the injections would enter the brain just two millimeters to the side of the periaqueductal grey, neither morphine nor the active brain extract were able to relieve pain.

Isolating It

In the course of six months Rabi devised a series of six relatively straightforward steps that purified the active material about 50,000 times from the original brain extract. He was able to show that the purified material comprised what seemed to be a single small peptide with about six amino acids. By late November of 1975 he knew the identity of these amino acids: tyrosine, glycine, phenylalanine, methionine, and leucine. There seemed to be more glycine than anything else.

Rabi now began to work out the sequence of the amino acids within the peptide. The procedure Rabi chose used a chemical reagent that chops off amino acids one at a time from only one side, the left-hand side, of the peptide. In this way he was able to determine that tyrosine was the leftmost amino acid. The fact that these initial forays showed a single amino acid at the left-hand side greatly increased our confidence that we were dealing with an absolutely pure peptide. If we had a mixture of peptides that had different amino acids at their left-hand side, we would have found two or more amino acids in the first experiments.

Isolation alone is quite an achievement. One can often purify a substance extensively, but obtaining it in absolutely pure form can often be extremely difficult, as minor impurities are very hard to eliminate. Confident that we had overcome these obstacles, we believed that completing the amino-acid sequence of what seemed to be a relatively small and simple peptide would pose no major problems. We guessed that it would take a month or so. In a little over six months Rabi had gone from a crude brain extract to isolation of an important biological substance and was close to finishing the identification of its chemical structure. Hughes and Kosterlitz had been working on the purification three or four times as long and, as far as we knew, they had not yet solved the problem.

Scooped

We were wrong. On the first or second day of December 1975, I received a short note from John Hughes in the mail and a xerox copy of a set of galley proofs from a manuscript entitled "Identification of Two Related Pentapeptides from the Brain with Potent Agonist Activity." The paper was to appear in the December 18 issue of *Nature*, the distinguished British scientific journal. Interestingly, as I read through the manuscript I didn't feel defeated or in any way saddened. Hughes and Kosterlitz's work on isolation of endogenous opiates was a model of scientific elegance. In all meetings and publications they had handled themselves with absolute integrity and fairness toward other workers in the field. Their now classic paper in *Nature*, a collaboration between Hughes, Kosterlitz, and the distinguished chemist Howard Morris, describes the results of astute detective work combined with the talent to take advantage of serendipitious events.[1]

When I had last learned the state of their work in the summer of 1975, the Scottish workers knew the identity of most of the amino acids within the opioid peptide. As it turns out, though most of the amino acids identified at that point were correct, there were a few errors. By September they had straightened out much of the amino-acid sequence. Linda Fothergill of the Biochemistry Department at Aberdeen had worked out for them the sequence of the first four amino acids, tyrosine-glycine-glycine-phenylalanine. Things had become muddy at this point. The researchers were not sure whether there was only a fifth amino acid or whether there might be a total of seven amino acids. Besides tyrosine, glycine, and phenylalanine, they had reasonable evidence for the existence of methionine and leucine and they thought that tryptophan might also be contained in the peptide. However, each time they conducted an experiment they got somewhat different results.

At this point Hughes journeyed to London to consult with Howard Morris. Morris is a mass spectrometrist who had just begun turning the power of the mass spectrometric technique to a study of peptide structure. A mass spectrometer is a large and rather expensive machine that works out the structures of chemicals based on their actual atomic masses. The periodic table of the elements that we memorize in high school chemistry gives the masses of elements in round numbers (oxygen is said to have an atomic weight of 16, carbon a weight of 12). However, the correct atomic masses can be determined with far greater precision, to three or even four decimal places. The mass spectrometer breaks a large chemical into fragments with a powerful electric beam. It sends the fragments through the field of a strong magnet and records the extent to which the various fragments of the molecule are deflected. Chemicals are deflected by the magnet in proportion to their atomic mass, so the machine provides a measure of the actual mass of each fragment. The researcher consults tables that indicate the masses of various elements and tries to figure out just which combination of elements gave rise to each fragment. Then by a sort of jigsaw-solving procedure the structure of the full molecule can be put together.

So in September or October of 1975, Howard Morris began to assault the pure peptide given to him by Hughes and Kosterlitz. In a very brief period of time he was able to explain their variable results. They had not been dealing with a single pure peptide at all but with a mixture of two related but distinct peptides. Each peptide consisted of five amino acids. The two peptides had the same first four amino acids, which explains why Linda Fothergill had no difficulty in working out the sequence of these four amino acids. The two peptides differed only in the amino acid at the fifth position which, for one of them was methionine, for the other leucine.

Rabi's findings fit well with the structure deduced by Morris. Rabi had found only a single amino acid, tyrosine, at the left-hand side, because the two peptides possessed the same left-most amino acid. Within eight weeks of the Hughes-Kosterlitz-Morris paper in *Nature,* Rabi had completed his sequencing of the opioid peptides. He came up with the same two peptide structures. Thus, 3,000 miles away, using a different assay technique, different purification procedures, and different animal species, we independently confirmed the Scottish results. There seemed little doubt of their validity.

I wanted to let Hughes and Kosterlitz know how much I admired their work and how happy I was for them. I sent them a telegram of congratulations, but this might not be quite enough. Paul Talalay, then chairman of Hopkins' Pharmacology Department, had grown up in England. Though he has lived in the United States since 1941, he has never shed his British accent and has remained an inveterate, incorrigible anglophile. He even maintains a charge account at Harrods, the distinguished London department store that sells anything anybody could possibly desire, and, of course, only the best of it all. I imposed on Paul's good will to have Harrods send Hughes and Kosterlitz a magnum of their finest cognac. When last I spoke with Hans, he was still working through the bottle, reserving it for special celebrations.

Naming the Baby

Back in May of 1975, almost seven months before the Hughes-Kosterlitz-Morris paper in *Nature,* opiate researchers had descended upon Airlie House in Virginia to caucus. Airlie House, located a few miles outside of Washington, D.C., is a resort designed specifically for group meetings. The sponsoring group went by the quaint designation, "The International Narcotic Research Club." It had been founded by Avram Goldstein as an informal, very small group of investigators doing basic research on opiate action. At the first meeting in 1972 there had been only about a dozen participants. The number had escalated a good bit at the second meeting in Chapel Hill in 1973, held only a few months after the first publications identifying the opiate receptor. The third meeting in Cocoyoc,

Mexico, in 1974, had comprised over a hundred researchers. By the time the group met at Airlie House in 1975 there were 200 or more, and many applicants had to be turned away for want of room.

Participants at that May meeting all presented their most recent findings in a number of areas, but the item of greatest interest was the progress being made in isolating the brain's own opiate-like peptide. Gavril presented the state of affairs in our laboratory. Terenius reported on his work. John Hughes described the advances in Aberdeen. The Scots were certainly well ahead of the rest of the field. At that point they had already purified the material to homogeneity, or certainly close to homogeneity, and had begun to analyze the composition of amino acids within the peptide. Their report stated, "Amino acid analysis after 6 M HC1 hydrolysis revealed the presence of glycine, methionine, phenylalanine, and tyrosine with no basic amino acids. A tentative estimation of molar ratios indicates three glycines, one phenylalanine, one methionine and one tyrosine with the probability of tryptophan or a derivative in unknown amounts." Clearly the only apparent error at that stage was the overestimation of glycine, the inclusion of tryptophan, and the absence of leucine.

Avram Goldstein reported on some new work from his laboratory that was rather different from everything else going on in the other groups. Goldstein and his students Hans Teschemacher and Brian Cox had been fascinated by reports that ACTH, the pituitary peptide hormone that stimulates the adrenal gland to release cortisol, influences the responses of rats to morphine. They purchased ACTH powder from a chemical supply house and tested its ability to mimic morphine's effects on contractions of intestinal muscle. To their delight, the powder produced morphine-like effects on the contractions of guinea pigs' intestines. Most importantly, those effects were blocked by the opiate antagonist naloxone. Since the relatively inexpensive ACTH preparation they had first purchased was not completely pure, they then obtained, at substantially greater expense, chemically pure ACTH. Pure ACTH was devoid of opiate-like activity. Apparently something other than ACTH in their first crude pituitary gland preparation had opiate-like activity. When they purified the material somewhat, it was possible to assess some of its chemical properties.

What they were studying was clearly different from what Hughes, Kosterlitz, Gavril, and I were investigating. Goldstein's material was much larger and displayed a number of sensitivities to enzymes that differed from the brain material in Scotland and Baltimore.

It was hard at that time for us to fathom why the pituitary gland should have any opiate-like material at all. From its location at the base of the brain, the pituitary gland discharges its secretions into the blood stream to influence glands all over the body. A peptide hormone released from the pituitary into the blood would hardly be expected to re-enter the brain.

Since all of the active groups working to isolate an opioid substance were present at the same time and in the same place, it was felt that we should talk about issues of common concern. One item for discussion was just what to call the material we were pursuing. Goldstein referred to his substance as POP, for pituitary opioid peptide. We used the term MLF, for morphine-like factor. Terenius also used the MLF designation. On the other hand, Hughes and Kosterlitz had already christened their molecule enkephalin, from a bastardization of a Greek term signifying "in the head." They did not wish to have any label denoting opiate-like, opioid, or morphine-like in the name. They reasoned that, while the molecule had been discovered for its opiate-like properties, this may represent only a minor aspect of its biological function. By keeping the name relatively noncommittal, they hoped that scientists would keep their minds open as to other possible functions.

It was extremely hot for May at Airlie House. The temperature and humidity were both in the 90s. I had played an exhausting set of tennis with Eric Simon and felt somewhat sun-strokey. I rested awhile and then went swimming and completely forgot about the caucus on nomenclature. However, Gavril, Candace, and Rabi participated.

According to them, at the meeting people went back and

forth with a variety of suggestions, none of which met with general agreement. Imagine a husband and wife attempting to name their forthcoming baby by gathering together both sets of grandparents-to-be along with all the uncles and a few cousins thrown in. After what seemed to be a hopeless stalemate, Eric Simon came up with a truly novel suggestion. "Why not call the substance 'endorphine,' which would be a contraction for endogenous morphine?" Everyone seemed to be agreeable except for Kosterlitz and Hughes, who grumbled about the various papers they had in press using the designation "enkephalin."

Avram Goldstein is a superb organizer. A few weeks after the Airlie House meeting, on June 20, he sent a round-robin letter soliciting comments and approval for the designation "endorphine." I agreed that we would use this name in all of our papers on the subject, several of which were already submitted to journals and several of which were in preparation.

Most recipients of the round-robin letter also agreed to go along with the endorphine terminology. Kosterlitz, however, responded, "It is obvious that there are different points of view with regard to the nomenclature of the morphine-like peptides. The standing of enkephalin is similar to that of heparin, serotonin, and gastrin; all these terms are indicative of the origin of biologically active substances without reference to structure, function, or concentration. Since there are a number of different peptides, it would be best to await the outcome of the work on their structures, actions, and functions before deciding on the possibility of a common name. Therefore, we suggest that enkephalin continue to be used for the Aberdeen and endorphine for the Palo Alto peptide."

To clarify, or to complicate, depending on one's point of view, the round-robin correspondence continued. On July 15 Goldstein sent out another group letter asking that the final *e* be deleted from endorphine: "I am pleased to hear from you that you all agree on Simon's proposed name endorphine. Actually, to conform to what seems to be an established tradition for biological amines nowadays, the name should terminate in 'in' rather than 'ine' so it would be *endorphin.*" Of the various biological amine neurotransmitters in the brain, serotonin does indeed terminate in "in." Somehow, Goldstein seemed to have forgotten that there are other amine neurotransmitters spelled rather differently, for example, norepinephrine and dopamine. Be that as it may, we all agreed to delete the *e*.

All of this correspondence took place in the summer of 1975. With the December 18 publication of Hughes and Kosterlitz's work, the disputations on names seemed moot. Since the Scots were the first to obtain the chemical structure of the brain's opioid peptides, they had the last laugh in the nomenclature battle. They designated the two peptides as enkephalins and distinguished them by referring to one as methionine-enkephalin and the other as leucine-enkephalin. I hastened to retrieve from publishers the galley proofs of several papers we had in press using the terminology "endorphin" and changed it to "enkephalin." I assumed that we had heard the last of "endorphin." Once again, I was wrong.

· 8 ·

Designer Drugs

The isolation of the enkephalins, like the discovery of the opiate receptor, spawned all sorts of new directions for research. There was one difference, however. The opiate receptor had sprung upon a scientific world that had given little thought to receptors and their conceptual implications. Thus, a substantial lag period ensued before scientists appreciated fully all the new insights that could emerge from opiate-receptor research. Even more important, researchers were slow to realize just how easy it was to measure opiate receptors and attack some of the major questions in drug and brain science.

In our small laboratory at Johns Hopkins, this lag was something of a blessing. For quite some time we had few competitors. Of course, we did not dawdle but proceeded on all fronts as quickly as possible. Still, it was relatively easy to skim the cream off the top of the opiate field.

With the enkephalins there was no such luck. By the beginning of 1976 the opiate-receptor concept had diffused widely throughout the scientific community. People well appreciated that the very existence of the opiate receptor portended a morphine-like neurotransmitter in the brain. In 1973 our competitors numbered only the small fraternity of pharmacologists with a special interest in narcotics. By 1976 some of the most eminent researchers in the world had moved in on the territory.

One can judge the new-found scientific respectability that

flowed to the opiate field simply by inspecting the credentials of researchers. Marshall Nirenberg, who won a Nobel Prize in 1968 for breaking the genetic code, developed cultures of nerve-like cells which can be grown by the millions in Petri dishes, much like bacteria, and which possess opiate receptors. The highest honor accorded an American scientist, short of a Nobel Prize, is membership in the National Academy of Sciences. In the early days of opiate receptor work, not a single opiate investigator was a member of the Academy. By four to five years after announcement of the opiate receptor, about a dozen National Academy members were involved in opiate-related investigations.

Basic scientists were not the only ones to seek the fruits of the opiate receptor and enkephalin discoveries. Most of the major pharmaceutical companies in the United States and Europe joined in the excitement. Their principal goal was to use the new findings in a quest for painkillers that might be less addicting and cause fewer side effects than conventional opiates.

Hope for Chronic Pain Relief

Months before the final structure of enkephalin was elucidated, pharmaceutical companies were already gearing up to develop pain-relieving drugs based on the structure of the brain's own morphine. The use of morphine and related opiates, the most powerful painkillers known, is restricted to instances in which the suffering is likely to be short-lived, such as post-operative pain. Conventional opiates are not administered on a routine basis to the millions of people suffering from chronic forms of pain, primarily because of the addictive potential of these drugs. People with damage to peripheral nerves sometimes develop exquisitely sharp pain that feels as though their flesh is being seared with a branding iron. A real branding iron burns only once, but patients with peripheral nerve damage suffer this excruciating pain hour after hour for years. Many commit suicide rather than face another day in such torment.

Less heinous, but far more common and often as debilitating, is the chronic pain associated with back and neck problems. Perhaps as many as one in four persons is afflicted with neck and back disabilities. Generally such disc problems arise when the gelatinous lubricant between vertebrae oozes out of its leatherlike covering and begins to compress spinal nerves. Victims are almost never comfortable in any position. The emotional turmoil that usually ensues worsens the perceived pain. Physicians often become frustrated and refer such individuals to psychiatrists, convinced that their patients cannot really be suffering that badly and must be imagining much of it.

Migraine and other forms of severe headache are just about as frequent as back and neck problems. Such headaches are often totally disabling, forcing the victim to lie in a darkened room for hours or days until an acute episode passes. Migraine headaches appear to produce their pain by the throbbing pulsations of dilated blood vessels in the head. Many drugs that influence blood vessels have been employed, with varying degrees of benefit, in dealing with migraine headaches. None is a panacea.

Potent opiates such as morphine could relieve the distress of these patients. But one cannot treat chronic pain with opiates, at least not with the opiates presently available. With chronic use the dose required to alleviate pain rapidly escalates. Soon the patient is as disabled from his or her opiate addiction as from the pain.

A nonaddicting opiate would be literally life-saving for millions of people with chronic pain. Fashioning a drug based on the brain's own opiate-like neurotransmitter sounded like an ideal direction for research in 1976. It seemed a good bet that an enkephalin-related drug would not be addicting—surely one could not become addicted to one's own neurotransmitter. Another reason for optimism in the drug industry was that synthesizing small peptides is one of the easiest tasks in organic chemistry. Dozens of derivatives can be turned out in a week's time. Surely a few months of fiddling with the molecule should result in an agent far more potent than the natural enkephalins and one custom-tailored to produce optimal pain relief.

Patent Disputes and Litigation

By late summer in 1975, Hughes and Kosterlitz knew the sequence of four out of the five amino acids in enkephalin. At that stage they did not know that the entire peptide was only five amino acids, so they did not appreciate just how close they were to the final solution. Furiously seeking advice from all available resources to decipher the last pieces of the puzzle, Hughes and Kosterlitz turned to peptide chemists in the drug industry. Two British companies they consulted with, Reckitt and Coleman and Burroughs-Wellcome, both had strong peptide-chemistry departments. From their discussions with the Scottish workers, chemists at these two corporations had a fairly substantial head start in the search for enkephalin derivatives. They commenced synthesizing enkephalin analogs simply by making educated guesses as to the identity of the amino acids beyond the first four.

The task was not without its technical problems, as it turned out. Being a peptide, enkephalin can be readily destroyed by any of a number of enzymes in the body. Since proteins are simply big peptides, our ability to digest proteins is dependent primarily upon the activity of a variety of these peptide-degrading enzymes, called peptidases. They attack all the chemical bonds that hold amino acids together in a peptide chain. The ability of the body's peptidases to degrade enkephalin became evident with the first administration of radioactively labeled enkephalins to animals. The enkephalins disappeared from the circulation within one minute.

The peptide chemist would have to find a means of protecting enkephalin derivatives from the actions of enzymes. The solution to this problem lay in the use of optical isomers. Amino acids, just like opiate drugs, can exist as stereoisomers, chemically identical molecules that are "right-handed" or "left-handed." Natural amino acids are always of the L-form, which rotates light to the left. For reasons that are not altogether clear, substituting one or more of the amino acids within a peptide with a D-form, the unnatural form, of the amino acid prevents peptidases from destroying most peptides. The resultant derivatives are thus quite stable in the body. Frequently, they are still active biologically.

Chemists at Burroughs-Wellcome and Reckitt and Coleman immediately proceeded to insert D-amino acids at various points in the enkephalin molecule and soon discovered that replacing glycine, which is the second amino acid in enkephalin, with the D-form of alanine accomplishes the desired goal. Glycine is such a small amino acid that it does not even have any optical isomers. Alanine is closely similar to glycine but is one carbon larger. The extra carbon in alanine confers optical activity on the molecule, so that there exist separate D and L forms of alanine. The chemists found that enkephalin derivatives with D-alanine at the number-2 position resist the actions of peptidases but still bind to opiate receptors. Indeed, most of these derivatives were just about as potent as enkephalin itself. It turned out that the D-alanine substitution was crucial for producing peptides that would resist breakdown but retain potency at the opiate receptor. Because of their advance information, Burroughs-Wellcome and Reckitt and Coleman were certainly among the first companies to obtain large numbers of enkephalin derivatives, mostly with the D-alanine substitution. But they were not alone. At the Sandoz Company in Basel, Switzerland, peptide chemists, among the best in the pharmaceutical industry, moved with extreme dispatch. Within one or two months they also discovered the D-alanine substitution and prepared numerous derivatives. Chemists at the famous Eli Lilly Drug Company in Indianapolis also independently identified the D-alanine substitution as a key to stabilizing enkephalin.

These chemical exploits soon eventuated in a dramatic legal confrontation. A chemical modification that improves the activity of a substance is definitely patentable. As more and more enkephalin derivatives were synthesized by drug companies, the importance of the D-alanine substitution and its protection by patent became apparent. Many of the enkephalin analogs incorporated D-alanine in position 2. Whoever held the D-alanine patent might be able to demand royalty payments from any drug company marketing an enkephalin derivative with the D-alanine substitution. If enkephalin-related drugs lived up to only a fraction of the potential that drug-industry analysts anticipated in 1976, then ownership of the D-alanine patent could be worth tens of millions of dollars.

Thus, it should come as no surprise that a battle for priority ensued. Ten parties, including the U.S. government, claimed to have discovered the D-alanine substitution first. My colleagues in the patent field tell me that an "interference" with so many participants was without precedent and represents a historic event in patent law.

As with so many court cases, the battle over D-alanine continued for several years. Finally, a judgment was handed down in the ten-party interference, and Sam Wilkinson of Burroughs-Wellcome was declared the winner.

Enkephalin Derivatives Fail

Despite all the scientific and legal brouhaha, it seems today that no analgesic drugs of therapeutic importance will emerge from the massive effort to synthesize enkephalin-related agents. The major pharmaceutical companies share this opinion. By 1980 Sandoz, which had assigned dozens of chemists, biochemists, and pharmacologists to the project, had dropped the enkephalin work. Even Burroughs-Wellcome, the ostensible winner of the race, has ceased efforts to obtain an enkephalin-related analgesic. Only the Eli Lilly Drug Company and Sandoz went as far as testing the analgesic effects of an enkephalin derivative in humans but with no dramatic result.

What went wrong? Never before in the pharmaceutical industry had so many companies worked so feverishly on a new drug development program. With that level of effort one usually witnesses several promising agents from each drug company involved. Programs of that magnitude usually continue for decades. The enkephalin drug development program may have set some sort of record for the rapidity of its demise.

Several elements seem to have been responsible. Stabilizing the enkephalin molecule turns out to have been the least of the problems to be solved. Even more important is developing an agent that will get into the brain. Since the brain is more sensitive to disruption than any other organ of the body, it has developed elaborate protective mechanisms. One of these is called the blood–brain barrier. Electrically charged molecules, which usually pose the greatest danger to brain function, are unable to penetrate the brain. Each amino acid possesses two charged groupings and enters the brain poorly. Peptides containing numerous amino acids do even worse. Very few peptides have ever been synthesized that pass readily into the brain. Chemists had hoped that they could modify the amino acids of enkephalin, making them very fat-like so that they could dissolve readily in the lipids of the brain despite their electric charges.

Sandoz came quite close. They synthesized an enkephalin derivative that incorporated D-alanine but also included modifications in several of the other amino acids of enkephalin. This agent was potent at opiate receptors and extremely stable metabolically. When injected directly into the brains of mice, it was 30,000 times more potent than enkephalin itself in relieving pain. Even more importantly, this derivative, designated FK-33824, produced analgesia in mice, rats, and monkeys when injected intravenously or subcutaneously.

Confident that they had attained the first effective enkephalin derivative, the Sandoz team published their findings in *Nature* and proceeded to clinical studies. FK-33824 was definitely active in human volunteers. In minute doses it produced a sense of heaviness in the extremities and a feeling of oppression in the chest, but no pain relief. Perhaps FK-33824 would relieve pain at higher doses. However, the peripheral effects were so unpleasant that higher doses could not be tested. The other Sandoz peptides seemed likely to suffer from the same limitations. Sandoz eventually closed up shop in the enkephalin field.

The saddest part of the whole enkephalin drug story is that it was based on what subsequently turned out to be a false premise—that one could not become addicted to one's own body chemicals. Surely, if enkephalin is a neurotransmitter, the brain's opiate receptors cannot become tolerant or else the brain's equilibrium would be destroyed. The enkephalin-forming neurons would be forced to turn out larger and larger amounts of enkephalin to overcome the insensitivity of the "tolerant" opiate receptors.

Drug-company scientists were certainly correct that the brain does not become tolerant to its own enkephalins. However, the story can be quite different for a stable enkephalin derivative. To understand the distinction, one must reflect on the ingredients that make for addiction at the receptor level. One does not become addicted as a result of taking a single dose of morphine. Pharmacologic research over the past fifty years has established that addiction takes place only if an opiate remains in continuous contact with opiate receptors at fairly high levels and for prolonged periods of time. The brain's neurotransmitters never stay in contact with their receptors for a prolonged period. All neurotransmitters are inactivated rapidly so that the receptors will be ready to welcome new neurotransmitter molecules. If neurotransmitters were not inactivated almost instantaneously, then synaptic transmission would be intolerably slow and inefficient. Our brains would simply stop operating.

Neurotransmitters can be inactivated by a variety of mechanisms. For peptide neurotransmitters, inactivation is mediated by peptidase enzymes that destroy neurotransmitter peptides. Thus, we do not become addicted to our own enkephalins because our opiate receptors never see them for longer than a millisecond. Enkephalin-degrading peptidases chop apart the enkephalin structure within instants of its activating the opiate receptor.

Of course, the hallmark of all the enkephalin drugs is that they are metabolically stable. In order to be useful as drugs, high levels of them must remain in contact with opiate receptors for a reasonably protracted interval. Not surprisingly, when addiction studies were done, it became clear that all the enkephalin drugs produced classic symptoms of opiate tolerance and withdrawal.

While drug-company chemists were busy synthesizing new enkephalin derivatives, other scientists began to conduct experiments to answer the simple question, Is enkephalin itself intrinsically addicting? Overcoming the problem of enkephalin's rapid degradation and its inability to enter the brain required some technical maneuvering. However, Dr. Larry Stein, a psychologist at the Wyeth Drug Company, was able to devise a relatively simple system in which fairly substantial doses of enkephalin were continuously infused directly into the brains of rats through a permanently implanted tube. Stein was able to demonstrate pain relief with enkephalins administered in this fashion. Moreover, he could also demonstrate tolerance and physical dependence. Stein's experiments clearly established that the efforts of the drug companies to develop a new panacea from enkephalin rested on shaky theoretical foundations.

Back to Basics

The failure of enkephalin to provide mega-bucks for the pharmaceutical industry was counterbalanced by magnificent successes in basic scientific research. Visualizing opiate receptors at a microscopic level had provided a giant step forward in understanding exactly how opiates exert their various pharmacologic actions. Obtaining microscopic images of enkephalin within the brain was similarly successful.

Besides his strong intellect, creativity, and hard work, Rabi Simantov displayed astonishing versatility. He could shift gears from one discipline to another with incredible ease. Rabi was able to pick up new techniques almost effortlessly. Thus, it came as no surprise to me when he volunteered to carry forward what seemed to be the next logical extension of enkephalin research.

To study a chemical in the body efficiently, a scientist must first be able to measure it. How might we measure enkephalin levels in tissues? Even before knowing the structure of enkephalin, Rabi had developed a relatively simple technique, designated the radioreceptor assay. Rabi merely took advantage of the procedure that we had used to monitor morphine-like material in brain extracts. He would measure enkephalin levels by determining the extent to which a tissue extract would compete with radioactive opiates for binding to the opiate receptor. Using this radioreceptor assay, Rabi had measured apparent enkephalin levels in many different regions of the monkey's brain before the structure of enkephalin was established. He had even measured the levels of morphine-like substance in many different species, ranging from man to the most primitive invertebrates.

The radioreceptor assay gave fairly reliable information. However, we always worried that some extraneous, nonspecific material in the tissue extracts might compete with the radioactive opiate for receptor binding, deceiving us into thinking that we were measuring enkephalin when we were, in fact, measuring something else. Clearly, we needed a more selective procedure for measuring enkephalin.

Radioimmunoassay is a classic procedure used to measure many biological chemicals and drugs. In many ways a radioimmunoassay is analogous to a radioreceptor assay, but it measures binding to antibodies instead of receptors. The major ingredient in a radioimmunoassay is the antibody to the substance that is to be measured. In this case, we would have to obtain antibodies to enkephalin. Then, we would measure the binding of radioactive enkephalin to its antibodies. Enkephalin concentrations in tissue extracts would be monitored on the basis of their ability to inhibit the binding of radioactive enkephalin to its antibodies.

Just about two weeks after the publication of Hughes and Kosterlitz's paper on the structure of enkephalin, our chemist friends at Sandoz mailed us vials containing synthetic met-enkephalin and synthetic leu-enkephalin. Rabi then immunized rabbits with the enkephalins, and in a relatively brief period of time he could show that rabbits had made antibodies against the two enkephalins. With active antibodies Rabi had no difficulty in developing a radioimmunoassay to measure tissue levels of enkephalin. We compared concentrations of enkephalin in different parts of the brain as measured by radioimmunoassay with our earlier results obtained by radioreceptor assay. Levels were essentially the same using either technique, ensuring that our earlier work with the radioreceptor assay had been valid.

Far more exciting things could be done with the enkephalin antibodies. At this time we had no concrete reason to assert that the enkephalins were, in fact, neurotransmitters. All we could say was that they were naturally occurring substances in the brain that had morphine-like effects at opiate receptors. Their home might not even be within neurons. For all we knew, they might exist predominantly in connective tissue cells in the brain. The first important criterion for a neurotransmitter is that it be found in specific populations of neurons.

Our enkephalin antibodies provided a means of determining whether enkephalin was contained within neuronal cells or other types of cells. The technique employed is called immunohistochemistry. Though Rabi had never done any neuroanatomy, he immediately began immunohistochemical studies to visualize enkephalin within cells in the brain. In collaboration with Michael Kuhar, who had considerable anatomical expertise, Rabi applied his enkephalin antibodies to thin brain slices on microscope slides. Presumably the antibodies would bind selectively to enkephalin contained within whatever cells normally stored it in the brain. How were we then to visualize the antibodies bound to enkephalin inside cells? Our antibodies had been raised in rabbits. We simply purchased from a chemical supply company antibodies raised in goats which were directed against the antibody-containing serum of guinea pigs or rabbits. There was one special modification of these antirabbit antibodies from goats. A fluorescent dye had been coupled to them. Thus, when we added the goat antibodies to the microscope slides, they would bind to the rabbit antibodies which were already bound to enkephalin. All we had to do was examine the slide under a fluorescent microscope. Wherever we saw fluorescence, we would be visualizing enkephalin-containing structures.

The microscopic pictures we obtained exceeded our expectations. First, enkephalin occurred exclusively in neurons. Taken together with everything else we knew about enkephalin by this time, this finding convinced me that we were dealing with a true opiate-like neurotransmitter.

Second, the places in which we found enkephalin neurons made perfect sense. In general, enkephalin neurons were found in the same parts of the brain where we had previously visualized opiate receptors at a microscopic level in our autoradiographic studies (see figure 8). Thus, the very dense band of opiate receptors in the substantia gelatinosa of the spinal cord was paralleled by a similarly dense concentration of enkephalin nerve endings in the same area. The periaqueductal grey, which had high concentrations of opiate receptors, was also rich in enkephalin neurons. The nucleus of the solitary tract displayed a remarkably high density of enkephalin nerve endings similar to its intense cluster of opiate receptors.

The extraordinary coincidence in the location of enkephalin neurons and opiate receptors established definitively that the enkephalins were the normally occurring substances whose function was to interact with opiate receptors. In other words, nature created opiate receptors to interact with their neurotransmitter enkephalin. Opiate receptors are, in fact, enkephalin receptors.

Subsequently, we and other groups were able to map enkephalin neurons throughout the brain in considerable detail. It was possible to localize the cell bodies of enkephalin neurons and follow the processes extending from the cells to synapses close by or far away. Indeed, mapping in detail the course of neuronal systems for a particular neurotransmitter is a powerful means of divining the function of that neurotransmitter in the brain. Besides enkephalin, we now know of sixty or more peptide neurotransmitters in the brain, each of which occurs in its own distinctive population of neurons. Localizing these peptides throughout the brain has been accomplished by numerous investigators, but most notably by Tomas Hokfelt of the Karolinska Institute in Stockholm, who pioneered the use of antibodies to visualize neurotransmitter peptides in the brain. While we were making antibodies to enkephalin

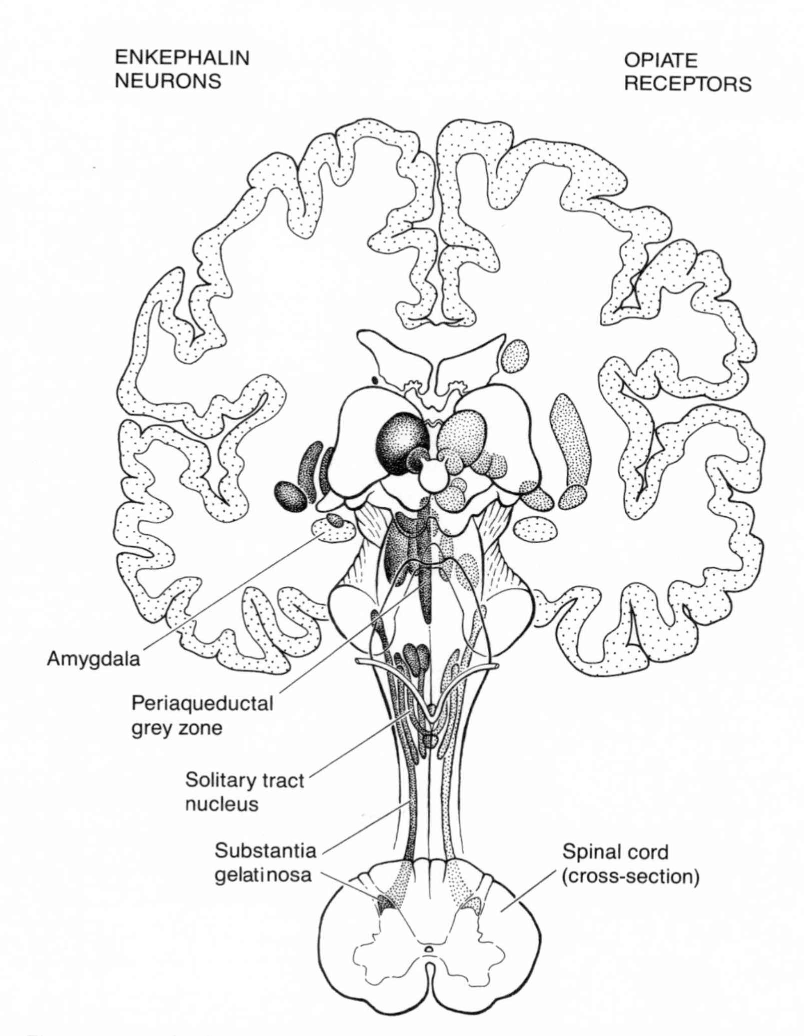

Figure 8. In the brain, enkephalin neurons (shown only on the left side in this cross-section) are found in the same locations as opiate receptors (shown here only on the right side). This discovery established convincingly that opiate receptors normally function as receptors for the neurotransmitter enkephalin.

and mapping enkephalin neurons, Hokfelt independently carried out a similar study and, in fact, published his findings first.

Resurrecting Endorphins

At the same time that drug companies were churning out enkephalin derivatives and researchers were localizing enkephalin neurons in the brain, a new group of opiate-like peptides was being characterized whose significance is unclear even at the present time.

The final, definitive elucidation of the enkephalin structure had been carried out by the chemist Howard Morris at Imperial College in London. Just about the time that he finished identifying the two structures of enkephalin, Morris took a break from his laboratory to sit in on an informal seminar given at Imperial College by a peptide biochemist, Derek Smyth. Smyth was working at the National Institute for Medical Research in Mill Hill, a suburb just north of London. Here he had been characterizing hormonal peptides within the pituitary. The pituitary gland produces the peptide hormones that are the master regulators of most of the glands of the body. The best-known pituitary peptides include growth hormone, thyroid stimulating hormone, several gonad stimulating hormones, and ACTH, the adrenal cortex stimulating hormone. Smyth chose to devote his efforts to studies of a pituitary peptide whose function was not at all clear.

Beta-lipotropin had been isolated in 1965 by the eminent peptide chemist C. H. Li of the University of California. Beta-lipotropin was one of the largest pituitary peptides known, comprising 91 amino acids. It was identified because of its ability to affect lipid metabolism by fat cells. However, this effect of beta-lipotropin was relatively weak, leading Derek Smyth and other researchers to speculate that the true function had nothing to do with fat tissue. Smyth reasoned that beta-lipotropin might be a precursor of other hormonal peptides that were cleaved from beta-lipotropin by peptidase enzymes. In his seminar Smyth pointed to various parts of the molecule where such cleavages might take place. One prime site for cleavage was between amino acids 60 and 61. Sitting a few rows back in the audience, Morris was astonished to note that embedded in the 91-amino-acid sequence of beta-lipotropin was a perfect copy of the 5 amino acids of methionine enkephalin at amino acids 61 through 65. Morris speculated that beta-lipotropin might function as a precursor of met-enkephalin. Indeed, when he sat down with Hughes and Kosterlitz to compose their classic paper on enkephalin isolation, Morris included this suggestion.

Smyth wondered whether other portions of the beta-lipotropin molecule might possess opiate-like activity. Since he was working on the chemistry of the large peptide and had already prepared a variety of fragments, Smyth was in a perfect position to test this theory. Working in the same building as Smyth were Nigel Birdsall and Ed Hulme, pharmacologists with receptor expertise. At Smyth's suggestion, they checked out the effect of various fragments of beta-lipotropin on opiate receptors. The fragment running from position 61 to the end, comprising almost a third of the beta-lipotropin molecule, can be readily digested away from beta-lipotropin by peptidase enzymes. Of the numerous fragments tested, this 61–91 fragment displayed the greatest opiate-receptor activity. Smyth and his colleagues worked expeditiously and were able to submit a manuscript to *Nature* on February 21, 1976, barely two months after publication of the enkephalin sequence.

Smyth was not the only one. Roger Guillemin of the Salk Institute in La Jolla, California, was there as well.

It's All Greek to Me

Guillemin had spent twenty years isolating peptide hormones that are secreted from part of the brain called the hypothalamus and then travel to the pituitary to regulate the release of pituitary peptide hormones. In the summer of 1975 Guillemin had already elucidated the structure of at least three such hypothalamic peptides, which regulate thyroid, growth, and gonadal hormones. The world of medicine agreed that these accomplishments were major landmarks. The salon conversation of medical researchers dealt often with the question of just how soon the Nobel Prize committee would recognize these accomplishments, which they did finally in 1977.

But Guillemin was restless. He was plotting strategies to isolate still other hypothalamic peptide hormones. He also was reading the emerging publications indicating that the brain must possess a morphine-like factor. At that time the enkephalin structure had not yet been elucidated, but it was clear that such a material existed. The techniques for measuring it were well established, and the Scottish group had already worked out some features of its chemical structure.

It occurred to Guillemin that his own laboratory was admirably equipped to attack such a problem, despite their lack of experience in the opiate field and despite their late start. In the process of isolating hypothalamic factors Guillemin had accumulated partially purified and concentrated extracts from the hypothalamus of hundreds of thousands of pigs. In retrospect, of great importance was the fact that most of the pituitary gland was included in the hypothalamic extracts. Guillemin was not overly concerned about the contamination. This might be a bonus, considering Avram Goldstein's perplexing observation that pituitary extracts contained opiate-like material.

Perhaps even more crucial than the abundant tissue supply was the peptide expertise of Guillemin's group. No one else had so much success working out complex structures of peptides isolated from brain tissue. Peptide chemists in his group also were skilled in synthesizing peptides. Thus, if they were to isolate a peptide from the brain or pituitary and establish its structure, they could confirm that the structure was correct by preparing a wholly synthetic version of the same peptide and showing that it had identical biological activity. Accordingly, sometime in the summer or fall of 1975 Guillemin and his colleagues Nicholas Ling and Roger Burgus began screening their preparations for morphine-like activity.

Their first task was to develop a means for evaluating opiate-like effects. Having no background themselves in the opiate field, Guillemin turned to Avram Goldstein. Goldstein explained how opiates inhibit electrically induced contractions of smooth muscle such as the guinea pig intestine. Guillemin set up this classical assay system in his own laboratory and proceeded to fractionate extracts of the mixture of pig hypothalamus and pituitary. In a remarkably brief period of time his group isolated a single peptide which accounted for the major morphine-like effects of the tissue extracts. They next established the amino-acid sequence. What they had isolated was a 16-amino-acid peptide whose first 5 amino acids were identical to met-enkephalin. Moreover, the sequence of amino acids represented numbers 61–76 of beta-lipotropin.

Eager to anoint their newly discovered opioid peptide with a unique name, they recalled the committee deliberations of the Narcotic Research Club in Airlie House and decided that the designation endorphin (without the *e*) would be ideal. By the time Guillemin wrote his paper, Hughes and Kosterlitz had published their complete isolation and sequencing of enkephalin. Guillemin felt that the term enkephalin should be reserved for the 5-amino-acid peptide of the Scottish workers. He seemed to have had a premonition that other opioid peptides might subsequently be recognized. Thus, he added the prefix alpha and named his 16-amino-acid peptide alpha-endorphin.

Guillemin communicated his initial manuscript on this work to *Compte Rendu Academie Science Paris* on January 26, 1976. This journal has an extremely short publication lag so that the article appeared in the February 23, 1976, issue. Thus, in a technical sense Guillemin had priority over Derek Smyth, whose paper had been submitted to *Nature* on February 21, 1976, and appeared in the April 29, 1976, issue. Of course, Smyth had isolated a different peptide from the pituitary, one which contained 31 amino acids, of which alpha-endorphin was only the first 16. The matter of pituitary opioid peptides was still very much unsettled.

Though well into his 60s in 1975, C. H. Li was hardly inactive. He directed a large group, the Hormone Research Laboratory, which carried forward the isolation of pituitary peptides and synthesis of derivatives. His team included sophisticated peptide chemists as well as physiologists capable of evaluating the activity of peptides. It is unclear just when Li commenced his experiments looking for opiate-like activity of pituitary peptides. However, his work must have received considerable impetus from the Hughes-Kosterlitz-Morris paper reporting that met-enkephalin represented a portion of beta-lipotropin. In any event, Li proceeded very rapidly to evaluate various pituitary fractions obtained in the purification of beta-lipotropin.

Like Guillemin, Li had no background in the opiate field. Also like Guillemin, Li turned for help to Avram Goldstein at Stanford, only 40 miles from Li's own laboratory in San Francisco. Goldstein agreed to examine pituitary fractions for their ability to compete for opiate-receptor binding.

Li found that a major component of his extracts contained substantial opiate-like activity. He showed that this component represented amino acids 61–91 of beta-lipotropin, a result identical to Derek Smyth's. Li named his material beta-endorphin, apparently because it derived from beta-lipotropin. Li communicated his publication to the *Proceedings of the National Academy of Sciences USA* on January 30, 1976, four days after Guillemin had submitted his paper for publication and three weeks earlier than Derek Smyth. While Avram Goldstein was not an author of Li's first paper on beta-endorphin, he was a coauthor in a paper published two months later in which beta-endorphin was shown to have opioid activity both in the guinea pig intestine and in opiate-receptor binding assays.

The naming of pituitary opioid peptides was rapidly becoming a confusing play on the Greek alphabet. Guillemin named his peptide alpha-endorphin presumably to indicate that it was the first of the endorphins to be identified. C. H. Li identified a different peptide simultaneously and designated it beta-endorphin, not because he felt it was second but because it derived from beta-lipotropin. Derek Smyth's independent and simultaneous isolation of beta-endorphin attracted somewhat less attention, perhaps because he did not label the substance as endorphin—a term that very quickly caught the scientific public's fancy—but as "C fragment," since it was a fragment obtained from the "C" region of beta-lipotropin that he had isolated before suspecting it of opiate-like activity. Soon thereafter, another fragment from this portion of the lipotropin molecule—one that included the met-enkephalin sequence—was isolated and shown to possess opioid activity. It was designated gamma-endorphin.

How many pituitary endorphins might exist? What was their relationship to enkephalin? Even more pertinently, what were they doing in the pituitary gland?

Each of the various investigators felt that the particular pituitary endorphin he had isolated was the one that mattered most. In those early months of 1976 the stakes seemed quite high. In lectures, some of the involved parties argued that alpha, beta, or gamma endorphins were the true opioid peptides of the brain and that enkephalin was only a breakdown product, produced accidentally during the isolation process. If this were the case, then Hughes and Kosterlitz would not receive credit for identifying the brain's own morphine. Instead, whoever found the correct molecule would merit the attendant rewards.

The dilemma of the Greek letters resolved itself fairly quickly. Careful isolation studies, especially those conducted by Derek Smyth, indicated that alpha-endorphin was an artifact of isolation procedures, as was gamma-endorphin. The major opioid peptide of the pituitary was beta-endorphin.

How beta-endorphin might relate to the enkephalins remained to be determined. Of course, for beta-endorphin and beta-lipotropin to be precursors of enkephalin presupposes that beta-endorphin and beta-lipotropin exist in the brain. By late 1976 techniques were developed to measure beta-endorphin by radioimmunoassay. Its concentrations in the pituitary gland were enormous, hundreds of times greater than the concentration of enkephalin in the brain. However, when comparable assays were employed with brain tissue, it became apparent that brain levels were quite low, only about one-tenth those of enkephalin. Moreover, beta-endorphin was not located in the same places as enkephalin. Floyd Bloom at the Salk Institute and Stanley Watson at Stanford visualized beta-endorphin-containing neurons with immunohistochemical techniques analogous to those that had imaged enkephalin neurons. Beta-endorphin was contained in a limited system of neurons in the brain, but these occurred in very different regions than enkephalin. Moreover, whereas enkephalin neurons and opiate receptors had almost identical locations, there were few spots in the brain where beta-endorphin neurons occurred in close proximity to opiate receptors. Clearly enkephalin must be the normally occurring neurotransmitter for opiate receptors. The function of beta-endorphin remained a puzzle.

Endorphins, Stress, and the Immune Response

Floyd Bloom once delivered a lecture which he titled, "The Bane of Pain Lies Mainly in the Brain." We feel pain in our brains; morphine acts in the brain (and to some extent in the spinal cord) to relieve pain. It seemed most improbable that the enormous amount of beta-endorphin in the pituitary gland could have anything to do with pain relief. Some researchers tried very hard to make it "fit," with conceptual contortions such as a proposal that beta-endorphin wiggles up the pituitary stalk into the brain and then meanders throughout the farthest corners of the entire brain seeking opiate receptors. However, the physical structure of the pituitary and the brain would make it extraordinarily difficult for beta-endorphin to circumnavigate such a tortuous route from the pituitary into and all over the brain. Why should Nature adopt such a convoluted pathway when enkephalin-containing neurons were already in intimate proximity to opiate receptors?

Other investigators suggested that beta-endorphin might be released from the pituitary directly into the general circulation, and circulating beta-endorphin would then enter the brain as a natural constituent of the blood, much like glucose. However, peptides penetrate poorly if at all from the blood into the brain, as the fiasco of the drug industry with its enkephalin drugs emphasized dramatically. Moreover, experiments injecting radioactive beta-endorphin into the blood stream showed clearly that little if any reached the brain.

Just what might beta-endorphin be doing in the body? All the accumulated evidence indicates that beta-endorphin in the pituitary gland is not there to regulate opiate receptors in the brain. It would seem more meaningful to seek a role in the body's periphery. Perhaps beta-endorphin is a pituitary peptide hormone just like the other pituitary hormones. Such a way of viewing matters follows from the 1977 efforts of two young investigators.

Dick Mains and Betty Eipper are a husband-and-wife research team. Working at the University of Colorado in Denver, they were trying to divine just how the pituitary gland synthesizes ACTH, the peptide hormone that travels to the adrenal glands to stimulate release of cortisol. They knew that peptides such as ACTH were cleaved from large protein precursors. Their task: to isolate the large ACTH precursor protein, pro-ACTH, and figure out just how pro-ACTH is converted to ACTH. They also were curious about the other peptide material within the pro-ACTH protein. Was it just an inert "carrier" for ACTH? It seemed strange that a large protein with a very detailed and specific amino-acid sequence would evolve, if much of it was not destined to convey any meaningful biological information.

While working out the biochemistry of the ACTH precursor, Mains and Eipper made a startling discovery. Contained within the sequence of pro-ACTH was beta-lipotropin, with beta-endorphin inside it. In the precursor molecule, beta-lipotropin was situated only two amino acids away from ACTH. One of the enzymes that normally cleaves hormonal peptides from their precursors could readily chop apart pro-ACTH to give rise to beta-lipotropin and ACTH. A subsequent enzymatic step sliced beta-endorphin out of the beta-lipotropin molecule.

What Mains and Eipper had discovered was an important biological principle—that Nature is quite economical. None of that large ACTH precursor seems to go to waste. The principle they enunciated was that a large precursor molecule may contain within it several distinct, biologically active peptides. Subsequent to their discovery, numerous researchers have found many precursors that contain multiple hormonal or neurotransmitter peptide products.

Scientists jumped quickly upon the Mains–Eipper breakthrough. Floyd Bloom and Stanley Watson used immunohistochemical techniques to show that the same cells which stain for ACTH do stain also for beta-endorphin. Roger Guillemin showed that the pituitary gland releases ACTH and beta-endorphin simultaneously and in response to the same stimuli. Since ACTH and beta-endorphin are literally part of the same precursor molecule, are stored in the same cells, and are released together, one might speculate about their having related functions. We know that ACTH is released in response to stress. It journeys through the blood to the adrenal gland, where it provokes the release of cortisol that in turn enables the body to combat stress. Since beta-endorphin is released at the same time as ACTH, perhaps it also influences the adrenal or some other peripheral organ to facilitate an appropriate response to threats in the environment.

One way to ascertain if a given tissue is the normal target for beta-endorphin would be to determine if that tissue possesses specific receptors for beta-endorphin. Pedro Cuatrecasas, a co-discoverer of the insulin receptor and thus an expert in peptide receptors, attacked this question. Together with his colleagues Kwen Chang and Eli Hazum, Cuatrecasas screened many peripheral tissues and finally found a group that seemed to have highly selective beta-endorphin receptors. These were the lymphocytes, the white blood cells that mediate immune responses. Interestingly, the beta-endorphin receptors on lymphocytes do not interact at all with opiates such as morphine. In other words, if the lymphocyte receptor is the true target of beta-endorphin activity, then beta-endorphin's normal body function is unrelated to its ability to stimulate opiate receptors. The fact that methionine enkephalin comprises part of the beta-endorphin molecule might turn out to be a misleading red herring.

What would lymphocytes, whose main job is to mediate

immunity, have to do with beta-endorphin, which we have supposed to be a stress-related hormone? No one knows for sure, but we can speculate. Immune responses are altered in stressful situations. Cortisol release from the adrenal during stress alters the immunologic responsiveness of lymphocytes. Perhaps beta-endorphin regulates the immune responses of lymphocytes. Perhaps drugs that influence these lymphocyte receptors would modify immunity in a therapeutic way. Perhaps such drugs would influence the body's immune rejection of transplanted hearts, kidneys, and livers. Such agents might even enable the body to reject tumor cells.

We do not yet know for certain that beta-endorphin has as its normal function the regulation of lymphocytes. However, this story teaches us to be prepared for the unexpected. Discoveries in one field more often than not first emanate from a disparate area. The creative innovator is ever alert.

· 9 ·

From Opiates to Schizophrenia

By early 1975, the strategies that Candace Pert and I had used to isolate the opiate receptor had been applied in my lab to identify receptors for several other neurotransmitters. Hank Yamamura had used one of the U.S. Army's chemical warfare agents (a potent acetylcholine-blocking drug, quinuclidinyl benzilate—QNB, for short) to identify the brain's major type of acetylcholine receptor. Anne Young had employed the drug strychnine, well known for its ability to cause convulsions, as a tool to label glycine receptors. Jim Bennett, an MD-PhD student, had used the psychedelic drug LSD (d-lysergic acid diethylamide) to measure receptors for serotonin, a neurotransmitter that regulates emotional behavior.

In all these instances the probe was a radioactive form of a drug that bound to the receptor, usually an antagonist. The neurotransmitters themselves, for some strange reason, had never worked out as probes to label receptors. Then, shortly before she left to begin her medical internship, Anne found an exception to this rule. Working with Steve Zukin, a medical student, she was able to measure receptors for GABA (gamma-aminobutyric acid), the major inhibitory neurotransmitter in the brain, with radioactively labeled GABA itself. Soon thereafter, Jim Bennett found that radioactively labeled serotonin was suitable for labeling serotonin receptors.

High on the "awareness index" of everyone in our lab was another neurotransmitter, dopamine. Since the chemical structure of dopamine resembles that of serotonin, Jim suggested that we might be able to label the dopamine receptor by simply using radioactive dopamine.

Dopamine was identified in the late 1950s in the cells of a few specific brain structures. One of these is the corpus striatum, the part of the brain that regulates motor activity. In Parkinson's disease—a disorder in which people are unable to move normally—the general function of the corpus striatum is abnormal. In 1960 Oleh Hornykiewicz, an Austrian researcher, found that levels of dopamine were depleted from the corpus striatum of patients with Parkinson's disease. Subsequently, George Cotzias, a Greek physician working in New York City, replaced the missing dopamine by treating patients with large doses of its amino-acid precursor, L-dopa. Symptoms of Parkinson's disease were dramatically alleviated. In other words, the symptoms of Parkinson's disease appear to be caused by a deficiency of dopamine in the corpus striatum. Relieving the dopamine deficiency with L-dopa, which is converted into dopamine in the corpus striatum, produces an almost miraculous reversal of symptoms. L-dopa is considered one of the breakthrough drugs in the second half of the twentieth century.

A second group of brain structures that have high concentrations of dopamine are the amygdala, the nucleus accumbens, the olfactory tubercle, and certain parts of the cerebral cortex. All are components of the limbic system, an ancient part of the brain that regulates emotional behavior.

Jim was too busy with other projects to do anything about dopamine receptors. So Sam Enna, a postdoctoral fellow, set up the appropriate experiments. They were immediately successful. Radioactive dopamine bound tightly to brain membranes. Its binding was greatest in parts of the brain that were known to possess the highest concentrations of dopamine. Other neurotransmitters were examined, but only dopamine interacted potently with these binding sites.

Next, we screened a number of drugs to find out which ones would compete with radioactive dopamine at the binding sites. The most potent competitors were a class of drugs known as the neuroleptics—mainstays in the treatment of schizophrenia. This came as no particular surprise. Dopamine had already been recognized as a possible mediator in the action of neuroleptic drugs. In fact, this link was the primary reason why labeling the dopamine receptor was so high on the list of priorities in my laboratory. Again and again in my research, I have found myself returning to themes of mental illness, and the disease that has held the greatest fascination since my first days of medical school is schizophrenia, the most debilitating of emotional disorders.

Nietzsche and the NIH

Most people go to medical school because they want to be doctors. Much of their medical school career is devoted to an obsessive mulling over the merits of this or that subspecialty. I had little interest in caring for people's diseased bodies when I became a medical student. The MD degree, for me, was just a means to an end: becoming a psychiatrist. What I really wanted to study was diseases of the mind.

In high school I had read a lot of philosophy, particularly the German philosopher Nietszche. I was taken with the poetic elegance of his writing, even in translation, as well as the radical hyperbole of his ideas. But his remarkable insights into the human psyche, especially the unconscious, were the most magnetic feature of Nietszche's writing, for me. It was but a small step from Nietszche to Freud. I consumed his works voraciously, and had even digested the massive three-volume biography by Ernest Jones, before entering college.

But what kind of profession is nineteenth-century philosophy for a nice twentieth-century Jewish boy? All my high school friends planned premedical majors in college. Though I had done reasonably well in high school sciences, none of them ever interested me. I had no aptitude for the craftwork of high school science labs, which I frankly detested. Still, plodding through some heavy science in college on the way to the required MD degree seemed a reasonable price to pay for a psychiatry residency training program that would lead to a lifetime contemplating the mind.

A summer job in a laboratory at the National Institutes of Health precipitated my first, and most important, scientific discovery—that basic medical research is not at all like the science one learns in high school or college. The important element in grown-up research is not technical virtuosity but original ideas. The best investigators are as creative as the most innovative artists and composers. Outstanding science demands the same kind of thinking processes as the most creative philosophy. So why not combine basic research with clinical psychiatry?

With this goal firmly in mind, during my first year in medical school I read everything I could get my hands on about schizophrenia, and I also read widely about the brain. I had the feeling that thinking deeply enough about the psychological nature of schizophrenic abnormalities might lead to insights into the associated abnormalities in brain function. Much of the impetus for this extracurricular work came from my professor of neurophysiology, Estelle Ramey. Estelle arranged for fellowship grants so that I could pursue some library research and organize my thoughts. What followed was my first publication, "A Mechanism of Schizophrenia"—a sophomorish attempt to mesh psychiatric and neurophysiologic findings into a model of brain function in schizophrenia.

Since then, my speculations about the neurobiological basis of schizophrenia have become more sophisticated and circumspect. But the fundamental urge to understand how this grave emotional and cognitive disorder is rooted in abnormalities of brain function has not left me. It was undoubtedly at work in 1975 when we turned our attention to the search for the dopamine receptor. To understand the connection between dopamine and schizophrenia requires some background information about the disease and the laboratory research that has attempted to demystify it.

Shattered Minds

Severe forms of schizophrenia affect about 1 percent of the world's population, and less serious variations may be 2–4 times as common. Since schizophrenia is a chronic disorder, which partially or totally incapacitates people for most of their adult life, its financial and social devastation may exceed that of more common and better-known conditions such as cancer and heart disease. Just the expense of in-patient care for schizophrenics in the United States averages $10–20 billion annually, while the lost productivity, unemployment, social security, and welfare payments may be 10–20 times that figure. Worse than the financial cost is the devastation that schizophrenic illness brings not only to the patient but also to family and friends. Whether patients are withdrawn or combative, everything about their lives is so irrational and out of keeping with convention that normal living is impossible for anyone dwelling in their social and emotional sphere.

A better understanding of schizophrenia would have payoffs that extend beyond patients and their families, however. The abnormalities of schizophrenic thinking and perceiving, more than any other disorders of the mind, might teach us fundamental truths about how the brain regulates emotional behavior. Hallucination is the most florid and easily recognized schizophrenic symptom. A hallucination is a perception of something that is not really there. Hallucinations can be visual or auditory, or even involve the senses of taste, smell, and touch. The most characteristic schizophrenic hallucinations are auditory. Sometimes the voices schizophrenics hear are friendly and supportive, but more often they are terrifying, telling the patient how vile a person he is, sometimes even ordering the patient to kill himself or somebody else. Since the voices are projections of the patient's own mind, hallucinations reflect what the patient thinks of himself and tell us much about the patient's pitiful self-image. Visual hallucinations, which are much less frequent, usually involve terrifying apparitions.

Delusions are false ideas. Typical schizophrenic delusions reflect persecution or grandiosity. One patient may feel that she is being followed constantly by the FBI; another may believe that he is the savior of mankind, Jesus Christ, Napoleon, or John Lennon.

While the existence of flagrant delusions and hallucinations makes the diagnosis of schizophrenia easy, one may be schizophrenic without such symptoms. The most characteristic feature of schizophrenia, present in all patients, is a unique type of thought disorder. This disturbance of thinking is apparent in conversations with patients. The patient may seem to be speaking in a logical discourse. However, one soon realizes that nothing really coheres. Individual sentences may seem plausible, but the ideas are connected to each other only in vague, tangential ways. Often one gradually senses that the patient does not even feel that his brain belongs to himself. He frequently senses that ideas are being implanted into his brain from without or that ideas are being broadcast from his brain. Thus, the patient may believe that you know just what he is thinking, even though he has not uttered a word.

Until the advent of neuroleptics in the mid-1950s, there was no effective drug treatment for schizophrenia. During the 1930s several physical treatments had been devised. Working in a small sanatorium in Berlin, Manfred Sakel had noticed that schizophrenic patients who were also diabetic and happened to be hypersensitive to insulin sometimes lost their psychotic symptoms temporarily when they went into a coma associated with low blood sugar. In 1933 Sakel began to systematically administer large doses of insulin to schizophrenics. In his procedure, more and more insulin was infused till the patient's blood sugar fell so low that he or she lapsed into coma.

In 1935 in Portugal, Egaz Moniz first severed the frontal lobes of the brain in schizophrenics. The drastic procedure of brain-tissue ablation was not altogether without scientific foundation. Moniz was aware of findings that damage to the frontal lobes of animals or humans from brain tumors or injuries was associated with lessened aggressive tendencies.

Also in the 1930s, there were informal rumors that epilepsy rarely occurred in schizophrenics. L. J. von Meduna, a Hungarian psychiatrist, also noticed that schizophrenics who, for one reason or another, had epileptic seizures often experienced a temporary respite from their symptoms. Accordingly, he initiated convulsive therapy, first with chemical convulsants and then with electrodes.

All of these therapies were quite popular well into the 1950s, as many psychiatrists felt strongly that they were beneficial. Their impact on the medical, scientific, and general public was so pronounced that the Nobel Prize in Physiology and Medicine was awarded in 1949 to Moniz for his work with frontal lobotomy. However, few controlled studies had been performed to assess whether the drama of these heroic procedures and the enthusiasm of the treating staff were the agents really responsible for producing symptomatic relief. Subsequent research showed that the rationale behind many of these treatments was at best shaky. Schizophrenia is no less common in epileptics than in nonepileptics. No real biological relationship between insulin, blood sugar, and schizophrenia has ever been demonstrated. As for frontal lobotomies, they do decrease aggressive behavior, but they also almost literally destroy an individual's personality. Moreover, aggression is not the fundamental problem in schizophrenia. Schizophrenics with frontal lobotomies may be more docile, but they are just as psychotic as before.

Psychiatrists today are embarrassed by these once-popular therapies. Some of them may have improved some patients' lives, but the benefits to patients were marginal at best. Yet the scientific advances that were made possible by these biological therapies were substantial. For example, psychological studies of the thousands of people who had frontal lobotomies have contributed to our appreciation of how this part of the brain influences behavior. And doses of histamine and steroid hormones administered to schizophrenics in 1949, at Creedmoor Hospital in New York, by Dr. Arthur Sackler and his colleagues helped to clarify how stress, allergy, and inflammation are regulated by the interactions of histamine, adrenal steroids, and adrenaline.

Then in 1952 the first of the modern antischizophrenic drugs, chlorpromazine, was introduced. This neuroleptic was a relatively weak antihistamine with sedative properties that had been used by the French neurosurgeon Henri Laborit as a preanesthetic medication for surgical patients.

Laborit was impressed with the tranquility of patients treated with chlorpromazine and suggested to his psychiatric colleagues that the drug might be of use in emotional disorders. The Parisian psychiatrists Jean Delay and Pierre Deniker were the first to evaluate chlorpromazine systematically and to identify clear therapeutic effects. They reported their initial findings in May 1952. Within two or three years, chlorpromazine was employed all over Europe and the United States. While it was useful in calming hyperactive manic patients, its ability to relieve schizophrenic symptoms was clearly unique. Carefully controlled studies comparing the effects of chlorpromazine and a placebo established that the drug acts upon the fundamental symptoms of schizophrenia. When new, less-sedating neuroleptics were developed and proved to be just as effective as chlorpromazine, it became apparent that relief of schizophrenic symptoms could not be explained simply by sedation. Indeed, numerous traditional sedatives such as phenobarbital are ineffective in treating schizophrenia.

Prior to the introduction of chlorpromazine, most patients with severe schizophrenia would spend almost all the rest of their days in a state mental hospital. A diagnosis of schizophrenia was much like being sentenced to a living death. The neuroleptic drugs changed all of that. Most patients could be discharged from the hospital and allowed to function in the community. The community mental health movement (sometimes called "deinstitutionalization") that commenced in the late 1950s was made possible by these drugs; even seriously ill schizophrenics could be at least partially rehabilitated. The population in state mental hospitals plummeted from a peak of about 550,000 in 1955 to only 150,000 in 1978.

Neuroleptics do not *cure* schizophrenia. In many patients whose symptoms have been greatly improved, there remains a vague peculiarity of thinking and feeling processes. Moreover, after neuroleptics are withdrawn, patients often relapse. Schizophrenics need their neuroleptics like diabetics require insulin, perhaps not for a lifetime but certainly for several years.

Though neuroleptics are not a panacea, their relatively selective effects on schizophrenic symptoms render them powerful tools for understanding the schizophrenic process. In other words, it seems reasonable to suppose that neuroleptics act in the brain at a site fairly closely related to fundamental abnormalities in schizophrenia. Accordingly, if we knew how these drugs act at a molecular level, we would know something relevant to the aberrant events in the brains of schizophrenics.

Molecular Models

How does one discern how a drug exerts its therapeutic actions? If some drugs in a class are clinically extremely potent, while others are inactive, and still others have intermediate activity, we can determine if their relative clinical potencies parallel their potencies in influencing a particular biochemical event. If they do, then that biochemical event is presumed to be a target of the drug.

Neuroleptics are chemically reactive substances that affect many biochemical processes. In 1962, the Swedish pharmacologist Arvid Carlsson measured the concentration of several neurotransmitters and their metabolic products in the brain after treating rats with several neuroleptics. He wondered if the metabolic pattern of any of these neurotransmitters would be affected by the different neuroleptic drugs in proportion to their therapeutic potency. Most of the neurotransmitters he measured flunked this potency test—that is, those that were weak or inactive in treating schizophrenia would produce just as much of the neurotransmitter or its metabolic product as the most therapeutically potent drugs. Dopamine was the one exception. Thus, something about dopamine metabolism was closely allied to the therapeutic actions of neuroleptics.

Why would a drug change the metabolic pattern of dopamine? Using a fairly convoluted line of reasoning, Carlsson guessed that neuroleptics had as their fundamental action a blockade of dopamine receptors. Following this blockade, the firing rate of various neurons would change, and ultimately one would find an alteration in the pattern of dopamine metabolic products. There was no way to prove or disprove the theory, since in 1962 there was no way to measure in test tubes receptors for dopamine or for any other neurotransmitter.

With so long-standing an interest in schizophrenia, I read the literature on neuroleptics and dopamine and thought much about diverse ramifications. One of the difficulties in Carlsson's theory was that the chemical structure of dopamine did not resemble very much the structure of any neuroleptics. If neuroleptics were blocking dopamine receptors, they should plug into the dopamine receptor as a key fits a lock.

In 1970 a new postdoctoral fellow joined my laboratory. Alan Horn had obtained his doctorate in organic chemistry at Cambridge University in England. He enjoyed making molecular models of drugs and wiggling them about, as children manipulate tinker toys, to fit the drugs into sites where they might exert their therapeutic effects. Alan was greatly inspired by the brilliant success of James Watson and Francis Crick, who discovered the structure of DNA by making molecular models, without performing a single experiment.

As Alan and I talked about the puzzle of dopamine and the neuroleptics, Alan pointed out that a drug like chlorpromazine can exist in an almost infinite number of shapes, since many of its chemical bonds are flexible and can twist or turn in any number of directions. Indeed, the number of potential shapes for the chlorpromazine molecule was so great that our chances of success seemed vanishingly small. There was one possible shortcut. When a chemical becomes part of a crystal, like sodium chloride in salt crystals, it loses all flexibility and assumes a rigid shape or conformation. Often the conformation of a chemical in crystal structure is similar to its shape when displaying biological activity. Alan suggested that we review the chemical literature to see if the crystal structure of chlorpromazine had been established.

Sure enough, it had. Alan and I built molecular models of chlorpromazine in its crystal conformation, and we did the same for dopamine. When we looked closely at these models, we saw clearly that a major portion of the chlorpromazine molecule was almost exactly superimposable upon dopamine (see figure 9). This explained quite nicely how chlorpromazine could block dopamine receptors.

This approach did not fulfill a cardinal criterion for explaining drug action, however. We had not shown that the relative "fit" of a series of neuroleptics correlated with their relative

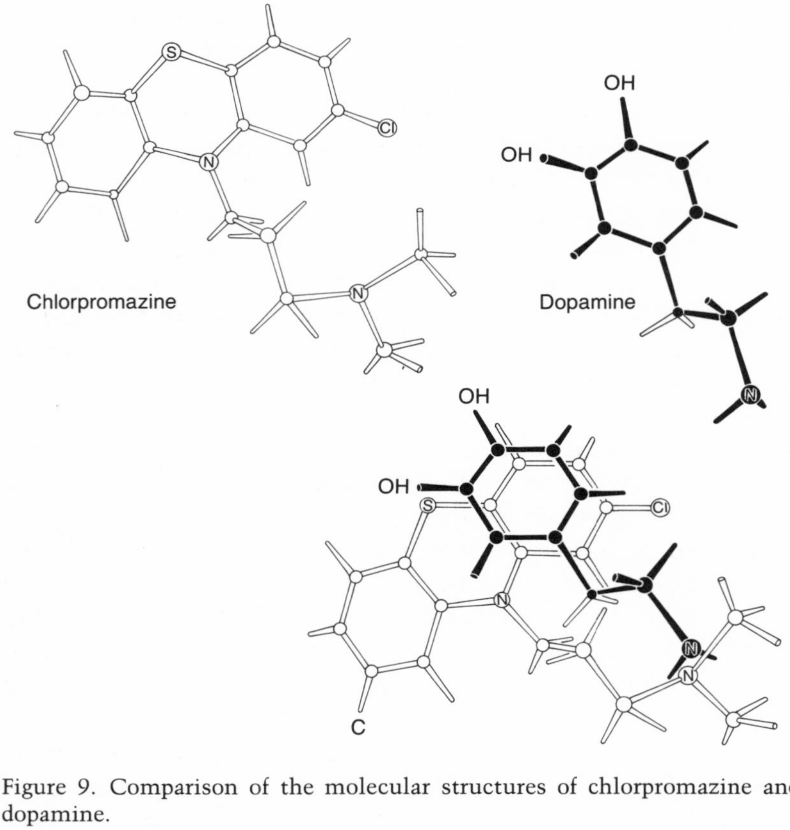

Figure 9. Comparison of the molecular structures of chlorpromazine and dopamine.

therapeutic potencies. We simply had no means of carrying out such an analysis. Numerous chlorpromazine derivatives could all assume the dopamine shape. However, with our simple stick models, we could not possibly measure the "potency" of a drug in assuming a particular shape.

About a year after Alan left Johns Hopkins, an opportunity to resolve this uncertainty arose when Andrew Feinberg walked into my laboratory. Unlike many medical students who are concrete in their thinking, eminently skilled at memorizing textbooks, but rather weak in abstraction, Andy had been a mathematical child prodigy. He entered Yale University when he was only thirteen and dazzled his professors. Yet in his senior year Andy shocked the mathematics faculty by proclaiming his intent to become a physician. He enrolled in Johns Hopkins Medical School and did reasonably well in the conventional medical school courses but felt a little out of place. He was eager to exercise his intellectual skills in some form of conceptual research. On his own, Andy performed mathematical calculations to reveal the structures of chemicals. A professor of biochemistry suggested that he visit with me, as my interest in the shape of drug molecules was well known.

Molecules do have different "potencies" in assuming particular shapes. In folding into a given shape, the molecule must consume a certain amount of energy. Presumably, assuming the dopamine shape will be more energetically favorable for some neuroleptics and less so for others. Calculating these energy states involved complex mathematics and extensive computer programs. Alan Horn and I lacked the mathematical expertise to tackle such a problem. For Andy Feinberg it was an amusing diversion from schoolwork.

In short order Andy developed computer programs to calculate the energetic states of neuroleptics in varying conformations. We were delighted to learn that the relative abilities of neuroleptics to fold into a dopamine-like shape fit magnificently with their relative potencies as antischizophrenic drugs.

Identifying Dopamine Receptors

Computer calculations and molecular models are fun, but they are no substitute for physically identifying the actual receptor molecule in the brain. All the models and calculations still left me unsatisfied. Our first biochemical attempts to assess the interactions of dopamine receptors with neuroleptics caused yet further concern.

After a neurotransmitter is recognized by its receptor site, there must be subsequent changes in the cell which tell it to fire more rapidly or more slowly. One such change is an opening or closing of a channel for the movement of ions, such as sodium or chloride. Another is an alteration in the synthesis of a second messenger molecule which, through further biochemical processing, alters overall cellular function. The best-known of all second messenger molecules is one called cyclic AMP. Some neurotransmitters seem to influence cell functioning by changing cellular levels of this chemical. Thus, measuring cyclic AMP might provide a biochemical probe, albeit indirect, to evaluate the receptor activity of neurotransmitters and drugs. In 1972 Paul Greengard, a brilliant biochemist and pharmacologist at Yale Medical School and one of the country's authorities on cyclic AMP, attempted to see whether dopamine affected levels of cyclic AMP in the brain. In homogenized corpus striatum, dopamine definitely increased levels of the chemical. Since dopamine exerted this effect far better than any other neurotransmitter tested, Greengard concluded that he now had a test-tube system to evaluate the activity of dopamine receptors. He tested a series of neuroleptics to determine their effect.

Paul found that chlorpromazine was fairly potent in blocking dopamine's effects on cyclic AMP. Chlorpromazine is a member of the chemical class known as phenothiazines. Paul found that the relative potencies of phenothiazine neuroleptics in blocking dopamine's effects on cyclic AMP paralleled their potency in controlling schizophrenic symptoms.

But the most potent neuroleptics are not phenothiazines at

all. They are from a chemical class called butyrophenones. Chemically, butyrophenones look very different from the phenothiazines. Haloperidol is a widely used butyrophenone neuroleptic that is roughly 10–20 times more potent at controlling schizophrenic symptoms than chlorpromazine. A relative of haloperidol called spiroperidol is the most potent of all neuroleptics. It is 10–20 times more active than haloperidol, hence hundreds of times more potent than chlorpromazine. Clearly, if neuroleptics act by blocking dopamine receptors, as monitored by Paul Greengard's system, then haloperidol and spiroperidol should be substantially more potent than chlorpromazine.

They aren't. In fact, in Greengard's experiments, spiroperidol and haloperidol were substantially weaker than chlorpromazine. Spiroperidol was among the weakest of all neuroleptics at blocking dopamine's effects on cyclic AMP. To get any effect with spiroperidol, Paul had to add it in concentrations thousands of times greater than occur at synapses in the brains of schizophrenics treated with therapeutic doses. Dopamine began to seem less and less relevant to schizophrenia.

It was just at this moment that Sam Enna completed the first group of successful experiments with radioactively labeled dopamine. Our confidence that dopamine played some critical role in the schizophrenic process began to soar once more. But both Sam and I could see that the project laying before us was immense. Sam had little free time to devote to dopamine. His major efforts were directed at characterizing the GABA receptor, a project he inherited from Anne Young. Fortunately, someone else was available.

David Burt was an impeccably trained biophysicist who had recently joined us as a postdoctoral fellow. His first pilot projects in the lab had failed, and he was looking for something promising. David had all the technical expertise and knowledge of biochemistry needed to analyze a receptor elegantly. His only limitation was a lack of background in pharmacology. David knew little about drugs and had hardly heard the word "neuroleptic." Yet, the major questions to be answered about the dopamine receptor required use of hundreds of neuroleptics whose variations were often subtle.

In contrast with David's background, Ian Creese, another postdoc, knew quite a lot about drugs and dopamine but little about biochemistry. Ian was trained as a psychologist and had done his doctoral research in Cambridge, England, evaluating dopamine-mediated behavior in rats. Publications from his doctoral work were so well recognized that in 1975, at age 26, Ian was regarded as one of the world's authorities on dopamine. Suspecting that future understanding of brain function would benefit more from biochemical than psychologic strategies, Ian had come to my lab to learn molecular approaches.

Ian and David made an ideal pair. Besides their complementary scientific backgrounds, the two were counterbalanced in temperament. David was an impeccable experimentalist and had much to teach Ian about biochemistry. David's careful, critical data analysis could counterbalance Ian's global brainstorming. Their collaboration worked out fairly much as predicted, a perfect match.

I was still leery about monitoring neurotransmitter receptors with the neurotransmitter itself. Consequently, I asked New England Nuclear Corporation to prepare a radioactively labeled batch of the neuroleptic haloperidol, which David and Ian compared with radioactive dopamine. Both substances successfully labeled dopamine receptors.

We evaluated all known neuroleptics for their potencies at the dopamine receptors labeled with radioactive dopamine and haloperidol. Within three weeks the task was complete. The potencies of a large number of neuroleptics in blocking dopamine receptors labeled with radioactive haloperidol paralleled closely their potencies in treating schizophrenic patients. For instance, spiroperidol, clinically the most potent neuroleptic, was also most potent in blocking dopamine receptors, in concentrations as low as one part per 10 billion. Haloperidol was also extremely potent, ten times more active than chlorpromazine. The closeness of the correlation was almost too good to believe, if one recalls that we were measuring drug effects at dopamine receptors in test-tube systems and then comparing them with clinical data derived from typical doses of neuroleptics used to treat patients. A drug's journey from a patient's mouth to receptors in his brain is a long one. Drugs differ considerably in their absorption from the stomach, metabolism in the liver, and penetration into the brain. Thus, two drugs might be identically potent as dopamine receptors in the test tube, but if one were metabolized 10 times more rapidly than another, then the first drug would seem to be only 1/10 as active as the second. Apparently, in our large series of neuroleptics all these factors cancelled one another out.

At last it was possible to state with a fairly high degree of confidence that the therapeutic actions of neuroleptics can be attributed to their ability to block dopamine receptors labeled with haloperidol. Our paper appeared in 1976; the conclusions still hold up today.

Not surprisingly, we were not alone in this quest for the dopamine receptor. Quite independently, Dr. Philip Seeman, a pharmacologist at the University of Toronto, also labeled dopamine receptors with radioactive haloperidol and dopamine. He, too, found that clinical potencies of neuroleptics correlate well with their ability to compete for dopamine receptors labeled with haloperidol. Interestingly, in contrast with the disputes about priorities in identifying opiate receptors, there was remarkably little in the way of arguments over who did what first with the dopamine receptor. Our two laboratories simply acknowledged the respective independent contributions of each, and that was the end of it.

Why were Seeman's group and our group so successful at correlating the therapeutic activity of neuroleptics with their ability to block dopamine receptors, when Paul Greengard had been unsuccessful? Greengard was surely measuring a form of dopamine receptor; yet the most potent neuroleptics, such as haloperidol and spiroperidol, were weak or inactive in his system. For several years we were unable to answer this question. Then work from several laboratories revealed the existence of at least two subtypes of dopamine receptors. In monitoring dopamine's effects on cyclic AMP, Greengard was measuring one subtype, which is referred to as D_1. Neuroleptics relieve schizophrenic symptoms by blocking another type of dopamine receptor, designated D_2, which is the type we were labeling with haloperidol. The function of the D_1 receptors is clearly not related to the therapeutic actions of neuroleptics.

Interestingly, the therapeutic effects of neuroleptics also did not parallel their potencies at binding sites labeled with radioactive dopamine itself. From our first experiments we had found several differences in the properties of dopamine receptors, depending on whether they were labeled with radioactive dopamine or haloperidol. For example, drugs such as spiroperidol and haloperidol were weak in competing with radioactive dopamine but potent in competing with radioactive haloperidol. The answer now was evident. Radioactive dopamine binds to the same D_1 receptors Greengard had been measuring via cyclic AMP effects, while radioactive haloperidol labels D_2 receptors. If D_1 receptors are not involved in neuroleptic actions, just what do they do? To this day no one knows, despite a decade of effort in over a hundred laboratories.

Amphetamine Psychosis

Neuroleptics do not merely quiet raving schizophrenics. They are not simply chemical straitjackets. Instead, these drugs seem to affect the fundamental thought disorders that characterize schizophrenia. It would seem reasonable that the site in the brain where the drug acts should be fairly close to the site where the fundamental aberration in schizophrenia occurs. Might dopamine have something to do with the schizophrenic process per se? If blocking the action of dopamine relieves schizophrenic symptoms, then one could speculate that schizophrenic abnormalities are related to excess dopamine release or perhaps hypersensitive dopamine receptors. Of course, the actual schizophrenic abnormality might not involve dopamine systems at all. The disorder might involve another system which, however, is neurally linked to dopamine. Whether directly or indirectly involved, dopamine seemed important for the disease.

Another group of drugs, the amphetamines, also provides clues that link dopamine to schizophrenia. Amphetamines were introduced in the late 1930s to energize fatigued patients. During World War II, German, British, and American armed forces used amphetamines to help soldiers and pilots function despite little sleep. Soon thereafter the appetite suppressant properties of amphetamines were appreciated, and the drugs became among the most widely prescribed agents in medicine.

By the late 1940s many people were abusing amphetamines, ingesting progressively larger doses as they developed tolerance. At high-enough doses, most users of amphetamines develop a psychosis that very much resembles acute paranoid schizophrenia. They have delusions that the police or the Mafia or some other threatening agent is after them, and they may arm themselves in self-defense or even injure innocent people whom they mistakenly regard as the enemy. Often they hear voices and display schizophrenic abnormalities in their thinking processes.

Many drugs can cause psychoses. Psychotomimetic drugs such as LSD by definition elicit psychosis. However, the psychosis that follows LSD ingestion is clearly unlike schizophrenia. Few psychiatrists will mistakenly label an individual under the influence of LSD as a schizophrenic. By contrast, many amphetamine users admitted to hospitals have been diagnosed as paranoid schizophrenic until the history of drug use was uncovered days or weeks later. In this sense, amphetamine psychosis is one of the best drug models of schizophrenia.

From the days that these drugs were first developed in the 1930s, psychiatrists have noted that low doses of amphetamines markedly exacerbate schizophrenic symptoms. The effect is quite selective, in that amphetamines do not seem to influence the symptoms of mania, depression, or anxiety. The prominent British psychiatrist William Sargent even administered a small dose of amphetamine to his patients as a test to verify a diagnosis of schizophrenia in cases where he was not confident of the diagnosis; if the symptoms worsened, he knew the disorder was schizophrenia. Research by numerous psychiatrists in succeeding years has confirmed Dr. Sargent's observations.

Dr. John Davis, a research psychiatrist at the University of Chicago, showed that amphetamines worsen schizophrenic symptoms regardless of the subtype of schizophrenia. The drugs aggravate whatever type of symptom the patient has been experiencing. Thus, paranoid patients become more paranoid. Nonparanoid schizophrenics—for instance, those with hebephrenic symptoms (incoherent speech and inappropriate, childish behavior) or catatonic symptoms (withdrawal, rigid immobility, muteness, and occasional excessive activity)—experience an increase in their own unique symptoms. The effects of amphetamines in already diagnosed schizophrenics seem more universally associated with the schizophrenic process than the induction of amphetamine psychosis in nonschizophrenics. All these psychosis-related effects of amphetamines suggest that something about amphetamine action in the brain is related to whatever neuronal abnormalities take place in schizophrenia. How amphetamines act at a molecular level might well be relevant to an understanding of schizophrenia.

What do amphetamines do to dopamine? Amphetamines act by almost literally pushing dopamine out of nerve endings in the brain. Thus, amphetamines increase the amount of dopamine released to act upon dopamine receptors. In this way, amphetamines produce an excessive stimulation of dopamine receptors, which leads to psychosis. One can titrate schizophrenic symptoms up or down by thus manipulating dopamine in the brain. More dopamine worsens schizophrenia; less dopamine relieves schizophrenic symptoms. Neuroleptics, on the other hand, block dopamine receptors and relieve psychotic symptoms. Not surprisingly, neuroleptics are the best antidotes for amphetamine psychosis. Neuroleptics rapidly and selectively "turn off" the psychotic symptoms produced by amphetamines.

Other evidence also supports the relationship of dopamine and schizophrenia. About the same time that chlorpromazine was found useful in treating schizophrenia, another drug was observed to have similar effects. Reserpine is extracted from the roots of an herb that has been used in India for many years. In Western medicine, reserpine is employed primarily to lower high blood pressure. However, folk tradition in India held that extracts of the reserpine-producing plant, *Rauwolfia serpentina*, could relieve psychotic symptoms. When reserpine was administered to schizophrenics in the early 1950s, patients' symptoms abated just as if they had received chlorpromazine. With the emergence of other neuroleptics, reserpine fell from favor, because of the markedly lowered blood pressure it elicited.

In the mid and late 1950s scientists found that reserpine impairs the ability of neurons in the brain to store certain neurotransmitters, including dopamine. In other words, reserpine literally depletes the brain of its dopamine. In this way it produces the same net result as the neuroleptics—less dopamine at receptors.

A Dopamine Concept of Schizophrenia

The concatenation of all these clues has suggested to researchers that schizophrenic symptoms might involve excessive dopamine activity. An alternative model would involve a fundamental disorder in schizophrenia at some other site in the brain, which indirectly elicits excessive dopamine activity. On the basis of studies done on schizophrenic brains obtained at autopsy, some scientists have reported abnormal levels of dopamine concentration in certain areas, while others have observed increased numbers of dopamine receptors. Both of these findings would fit in quite nicely with a dopamine model of schizophrenia.

There are reasons to reserve judgement, however. Since almost all schizophrenics are treated with neuroleptics, most of the brain samples analyzed are from patients who had received large doses of neuroleptics for many years. Continuous treatment of rats with high doses of neuroleptics produces biochemical changes closely resembling the alterations observed in the autopsied brains of schizophrenics. The abnormalities in levels of dopamine or in numbers of dopamine receptors may be more the product of drug treatment than of schizophrenia (see figure 10). In one study in our laboratory, Ian Creese, collaborating with the British investigators Leslie Iversen and Angus McKay, found increased numbers of dopamine receptors only in the brains of schizophrenics who had received large doses of neuroleptic drugs. Schizophrenics treated with lower doses for shorter periods of time had normal numbers of dopamine receptors.

One way to resolve many of these contradictions would be to measure dopamine receptors in living patients. Some recent developments make such a seemingly impossible study fea-

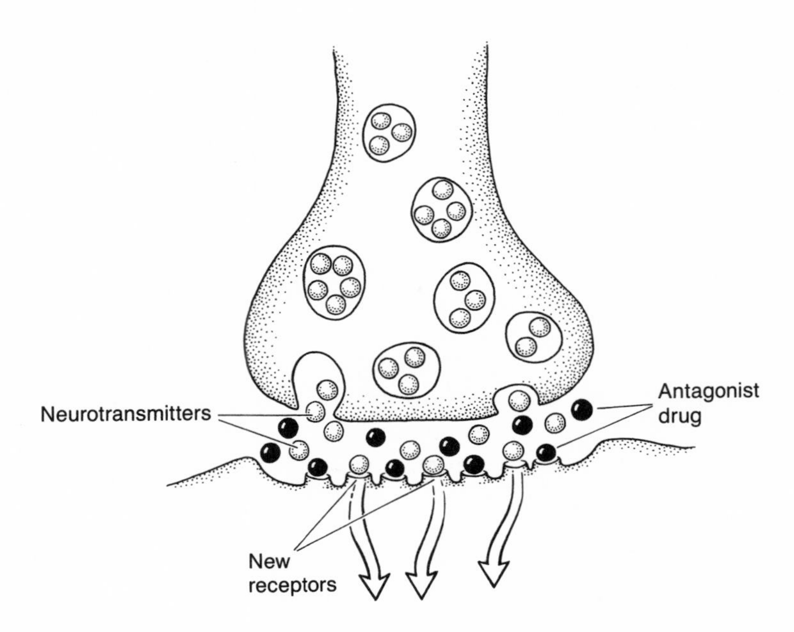

Figure 10. When neurotransmitter receptors are blocked by antagonist drugs such as the neuroleptics, the brain seems to try to overcome the blockade by forming new receptors.

sible. One of the modern miracles of diagnostic medicine is positron emission tomography—PET scanning for short. PET scanning provides a unique tool for visualizing radioactive chemicals within the human body, especially the brain. At Johns Hopkins, the PET scanner has been largely used to visualize neurotransmitter receptors, especially the dopamine receptor. In principle, the approach is quite simple. When Ian Creese measured dopamine receptors in our laboratory, he monitored the binding of radioactive neuroleptics to brain membranes in test tubes. It was subsequently shown that one can label dopamine receptors in intact animals. If a particular radiolabeled neuroleptic is injected intravenously, essentially all the radioactivity in the corpus striatum of the rat's brain is bound to dopamine receptors. A group of Johns Hopkins researchers, led by Dr. Henry Wagner, intravenously injected schizophrenic subjects with a derivative of the neuroleptic spiroperidol that had been labeled with the isotope carbon-11. Carbon-11 emits positrons that can be detected by the PET scanner camera. These human subjects, just like the rats, showed an intense labeling of dopamine receptors in the corpus striatum. It is even possible to visualize the dopamine receptors in the limbic system.

Schizophrenics have now been compared with nonschizophrenics. In the Hopkins experiments, numbers of dopamine receptors appear to be increased in schizophrenia. However, comparable experiments in Stockholm show no similar abnormalities among schizophrenics. Thus, the jury is still out on this knotty yet very important question.

Locating Schizophrenia in the Brain

Where in the brain do neuroleptics block dopamine receptors to relieve schizophrenic symptoms? Where in the brain do amphetamines release dopamine to worsen schizophrenic symptoms and cause amphetamine psychosis? Which of the several discrete dopamine neuronal systems in the brain is responsible for these drug actions?

Using a variety of neuroanatomical techniques, scientists

have been able to map out the location of the various dopamine neuronal groups. As I mentioned earlier, there are two major dopamine pathways in the brain (see figure 11). One of them sends its neuronal projections to the corpus striatum, which regulates body movements; this pathway degenerates in Parkinson's disease. By blocking dopamine receptors in the corpus striatum of schizophrenics, neuroleptic drugs create a functional dopamine deficiency. Among the major side effects of neuroleptic drugs are movement abnormalities that resemble Parkinson's disease. Another group of dopamine neurons projects to the limbic system of the brain, which regulates emotional behavior. Most researchers assume that the antischizophrenic actions of neuroleptics derive from blockade of these limbic dopamine receptors. Presumably, amphetamines worsen schizophrenic symptoms by releasing dopamine from neurons in the limbic system.

Numerous structures in the limbic system abound in dopamine neurons. One is the amygdala, whose stimulation in animals causes fierce rage responses. Cats stimulated this way sometimes behave as if they are searching for an imaginary enemy. There are also major dopamine projections to the olfactory tubercle. This part of the limbic system is closely associated with the olfactory bulb, which is the first way station for receiving information about smell. In most subhuman species, the sense of smell is crucial. Sexual behavior in animals is regulated by olfactory signals; friends and enemies are discriminated by their characteristic odor. In humans, these olfactory structures seem more related to emotional behavior but deal with the same life-or-death fear and anger responses that are elicited by odors in animals.

One dopamine system projects to the frontal lobes of the cerebral cortex, especially to those that developed earliest in evolution. These most ancient parts of the cerebral cortex are also involved in emotional responses. In performing frontal lobotomies in schizophrenics, neurosurgeons passed a metal wire through the frontal lobes at about the level of entrance of the dopamine neurons. Though frontal lobotomies rarely

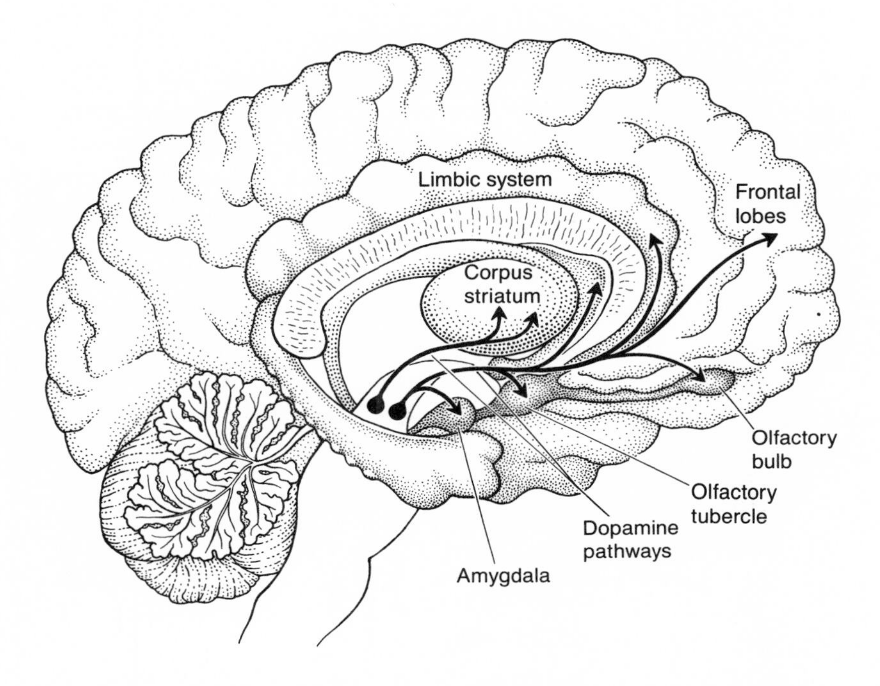

Figure 11. Dopamine neurons project to different areas of the brain that seem to be involved in two distinct diseases—schizophrenia and Parkinson's disease. It is well established that Parkinson's disease—a disorder characterized by peculiarities of movement and posture—results from a deficit of dopamine in the corpus striatum. Some evidence suggests that schizophrenia entails an excess of dopamine effects, particularly in the limbic system and frontal lobes. Neuroleptic drugs lessen the symptoms of schizophrenia by blocking dopamine receptors in the limbic system and other parts of the brain that mediate emotion. But because neuroleptics also bind to receptors in the corpus striatum and block their access to dopamine, a side effect of neuroleptic drug treatment for schizophrenia is a condition that strongly resembles Parkinson's disease.

reversed schizophrenic symptoms fully, they often provided some improvement.

Certainly the portions of the limbic system that receive dopamine projections have characteristics that would fit well with a role for dopamine in schizophrenia. Till now it has not been clear how one might elucidate which of these systems is involved in the actions of drugs in schizophrenia and perhaps in the schizophrenic process itself. New technologies, such as PET scanning, may provide the needed probes. In the not-too-distant future scientists may learn just where in the brain lie the abnormalities that determine schizophrenia.

· 10 ·

Receptivity

Before the advent of Valium and Librium, most physicians never dreamed that drugs might treat anxiety. Psychoanalysis spoke often of anxiety and its role in neuroses, but physicians did not appreciate clearly that anxiety is as biological and potentially treatable as a cough or runny nose.

Many serious emotional disabilities are still not adequately recognized as potential targets for drug therapy. For instance, extreme, morbid shyness greatly impairs the day-to-day functioning of many people. Diagnostic manuals label such individuals as schizoid personality. Often therapists offer psychodynamic explanations for this inability to relate to other human beings, and sometimes psychoanalysts believe they have traced the difficulty to family interactions in earliest childhood. However, physicians and therapists rarely recognize this behavior as a target symptom, like depression or anxiety, that is amenable to therapeutic attack with drugs.

Alcoholism and other forms of drug addiction are hardly treated at all by pharmaceutical agents. Yet there is abundant evidence that something is biologically aberrant in many alcoholics. Familial studies indicate that a predisposition toward alcoholism is determined by genetic factors. The genetic "load" for alcoholism is often as great as it is for schizophrenia or depression. If there is a biological underpinning for some forms of alcoholism, then it is reasonable to hope for effective drug intervention. Recent research indicates that alcohol, barbiturates, and related drugs act via the neurotransmitter GABA.

Thus, we already have hints for directions to go in seeking drug treatment in alcoholism.

Even among the emotional debilities where a biological cause is well recognized, there remains much to be done in the way of better drug treatment. Though excellent antidepressant drugs are available, none is perfect. All the antidepressants on the market today must be taken for about two weeks before a therapeutic effect becomes evident. Anyone close to a seriously depressed person knows that those two weeks can be a living hell. The risk of suicide is particularly high during this lag period before a patient's mood improves. Psychiatrists beg the drug companies to deliver an agent that will act immediately. With antidepressants, as with neuroleptics, the depression often returns when the drug is stopped. Yet maintaining patients on substantial doses of antidepressants or neuroleptics for years is a most unwelcome prospect, since all of these drugs elicit side effects that can become progressively worse with long-term treatment. So with depression as with schizophrenia, we need drugs that act "surgically" to abort the illness without risk of recurrence.

Dawning of the Receptor Era

Before the advent of receptor research, drug development was quite empirical and crude. Chemists and pharmacologists in the large drug companies had little to do with each other. Teams of well-trained chemists would synthesize thousands of new chemical entities each year, based largely on their own particular interest in specific subdivisions of chemistry, with very little regard for therapeutic application. Every week dozens of chemicals would then be turned over to pharmacologists to screen for therapeutic indications. Since there was no way to predict the therapeutic action, if any, of a given chemical, each one would be screened in many test procedures. For instance, a screen for anticonvulsant drugs might go something like this: Wires would be twisted around the heads of rats and then hooked to an electric generator. With appropriate stimulation, rats convulsed. Groups of rats were then treated with test chemicals. If any chemical reduced the number of rats convulsing, it was regarded as a candidate for the treatment of epilepsy.

Very little profound scientific thinking is involved in such a process. Indeed, during the years that drug companies were largely using this brute-force approach to drug evaluation, many biochemists and pharmacologists chose to leave science altogether rather than work for a drug company. Such a screening process was inefficient—as many as ten thousand compounds would have to be screened before one was found with enough promise to warrant introduction into human subjects.

Things are changing in the art and science of drug discovery. The introduction of modern molecular biology, of which receptor research is but one example, has greatly altered the research strategy of pharmaceutical companies. For instance, if a drug company's scientists wish to develop a new antischizophrenic agent, they might approach the problem this way: Knowing that effective neuroleptic-antischizophrenic agents block dopamine receptors, they evaluate compounds initially in test tubes for their potency in blocking dopamine receptors. Thus, the dopamine receptor will be the primary screen. It can be a most efficient procedure. A single technician can run through 500 test tubes in a day; some drug companies have developed automated receptor-binding machinery that works twenty-four hours a day. Even an extensive evaluation at receptor sites requires no more than a thousandth of a gram of the test chemical, which can usually be synthesized in a few days. By contrast, screening tests in intact animals often consume 50 or 100 grams, whose synthesis requires an extra month of scaling up.

Test-tube screening of drugs provides other advantages. Once a chemical is positive in a test, the drug company synthesizes dozens of variants of the molecule in an effort to further enhance potency and selectivity. Such efforts to increase the effectiveness of drugs in small doses is extremely important. If a drug can exert its therapeutic action at an extremely low dose, then it is less likely to produce side effects at sites of action that are different from the target receptor for therapeutic effects. If compounds are evaluated only in intact animals, then identifying chemical alterations that systematically increase activity is extraordinarily difficult. For instance, in a series of five chemicals of differing potencies in intact animals, one agent may be more active because it is not degraded in the liver, another may be more potent because it is absorbed better from the intestines, a third may be more active because it penetrates from the blood into the target organ more efficiently, while a fourth might actually have greater affinity for receptor sites. How is a chemist to use the potency data from these animal tests to improve activity? By contrast, in evaluating molecular effects on receptors, one can define a precise potency for each compound. The ways in which chemical changes in the molecule influence receptor affinity are thus readily apparent, enabling the chemist to move efficiently toward more potent and selective agents.

Screening chemicals for their effects on dopamine receptors is now carried out routinely by the drug industry in developing antischizophrenic drugs. Receptor technology can also be used to identify potential side effects. Some neuroleptic drugs are extremely sedating. For a young person on neuroleptics who is trying to get through college, sedation can be intolerable. Once it was possible to measure receptors for norepinephrine, we found that the relative sedative potential of neuroleptic drugs could be predicted by their ability to block one subtype of norepinephrine receptor. One can screen for drug effects on these receptors just as readily as one can measure dopamine receptors. Thus, if a drug company wishes to develop a potent but nonsedating neuroleptic, their pharmacologists screen for the effects of chemicals both on dopamine and on this specific subtype of norepinephrine receptors.

On the Horizon

Even more exciting than improving older drugs is the prospect of developing completely new classes of drugs that act upon some of the most recently discovered neurotransmitter systems in the brain. Prior to the identification of the enkephalins, a few peptides had been suggested as possible neurotransmitters, but relatively few scientists were involved in research on peptides as neurotransmitters. The immense scientific and public attention that attended the discovery of the enkephalins changed all of that. Within a few years, numerous other peptide neurotransmitters were identified. Almost every month or two a new peptide transmitter candidate is reported in the scientific literature. At the beginning of 1989 the number of peptide neurotransmitters was somewhere between 60 and 70; the discovery process shows no signs of abating.

Each of the new neurotransmitter peptides has a unique pattern of distribution in specific neurons in the brain. In most instances these distributions are just as interesting as those of the enkephalins. Some peptides are highly localized in parts of the cerebral cortex that may be involved in the highest intellectual functions. Others are concentrated in areas of the brain that regulate processes such as blood pressure and heart rate. Nobody knows for certain just what the function is of each of these new neurotransmitters.

Drugs—the most powerful tool in teaching us how neurotransmitters regulate brain function—do not yet exist that modulate the activity of these newly discovered neurotransmitters. However, there is every reason to believe that drugs which selectively stimulate or block their receptors, or influence their formation and destruction, will soon be just as important in the treatment of mental and emotional disorders as current antischizophrenic, antidepressant, antianxiety, anticonvulsant, and other centrally active agents have been. Indeed, the ability to use modern molecular strategies to fashion extremely potent and selective agents suggests that a whole new generation of more useful drugs is about to be born. New drugs might play a useful role in influencing emotions and behaviors in ways we cannot even imagine today.

Besides the practical benefits of new drugs, receptor research may lead us to a fundamental understanding of just what has gone wrong in the brains of patients with psychiatric disease. The technique of positron emission tomography (PET scanning) has already made it possible to visualize dopamine and opiate receptors in intact human beings. It is quite likely that receptor sites for many other transmitters will be imaged in a few years. Many scientists have theorized that particular diseases involve abnormalities in receptors. Carefully examining the numbers of receptors for each of these transmitters in different parts of the brain might identify such abnormalities.

In the nineteenth century almost half the psychotics in mental hospitals suffered from either "general paresis" or pellagra. The discovery that general paresis was an advanced stage of syphilis and that syphilis could be virtually eradicated by penicillin was a most important breakthrough in psychiatry. Similarly, the finding that pellagra stems from vitamin B deficiency led to curative treatment. Since the early twentieth century few inroads have been made into the causes of the other major mental illnesses. Indeed, psychiatry textbooks presently label schizophrenia and manic-depressive illness as "functional" psychoses, to distinguish them from "organic" psychoses such as those caused by syphilis or vitamin deficiency. The term "functional" suggests that no specific biochemical abnormality exists and reflects a discouraged ennui, a reluctance even to search for causes. The new molecular approaches to the brain promise to change this state of affairs, and the change may be much sooner than any of us expect.

Who Will Make the Discoveries?

From encounters with outstanding scientists over the years, I have learned that there is no single road to success when it comes to scientific discovery. Some Nobel laureates are sloppy and intuitive; others are impeccable and precise. Yet scholars of the discovery process know that 90–99 percent of the most important scientific findings are made by 1–10 percent of scientists. A small, gifted group of investigators seems invariably to spawn creative ideas and crucial observations. What distinguishes these innovative few?

Historians, sociologists, and philosophers of science have

tried for years to identify qualities that make for outstanding scientific contributions. Many of these studies have focused on the distinguishing characteristics of Nobel laureates. What they found was that intelligence as measured by IQ scores, schools attended, socioeconomic status, parental occupation, the warmth or coldness of childhood environment—none of these factors seem to be particularly relevant. The best way to predict who will make a discovery worthy of a Nobel Prize is simply to examine who trained whom.

Most scientists are intelligent but not remarkably brilliant. Certainly, a good scientist should be incisive, capable of separating artifacts from meaningful data. But what seems to matter most in the giants is a creativity mixed together with these critical skills. And from my own observations I am convinced that this *creative* aspect of scientific discovery is best communicated through apprenticeship. In this respect science is no different from other creative endeavors. Among musicians and artists, the chain of mentor–student relationships embraces all the greatest innovators. Throughout the Renaissance the most important painters followed one another in student–teacher pairs for several generations. Chains of teachers and their students can be similarly traced among leading composers and architects.

Science, like other aspects of a culture, is a communal activity. The way in which scientists work together in teams is crucial to the discovery process. And the most important team of all is the one made up of a mentor and a student. Just how do mentors influence their laboratory students? I can speak only from my own experience. I spent two years with Julius Axelrod in a small laboratory immediately adjacent to his. We would talk about research several times every day. I soaked him up through all the pores of my being.

Julie was more concerned with dreaming up new ideas than with the technical virtuosity of experimentation. In fact, both of us are a little sloppy and definitely impatient. In his eagerness to acquire new data, Axelrod was ever devising simpler ways of conducting experiments. This seemingly trivial quirk was one of the more important gifts I received from him. In science, most advances come from new methods for measuring some aspect of the natural world. Axelrod's lack of patience with tedious textbook methodologies led him to develop novel, simple, yet sensitive and specific means of measuring the activity of enzymes and the disposition of neurotransmitters. New methodologies played a key role in his Nobel-Prize-winning discoveries.

Technical expertise is no more my forte than it is Axelrod's. I remember well the first week of the first summer that I worked as a college student at NIH. I neglected to tighten the top of a rotor, the spinning part, of a centrifuge that cost $15,000 in 1958. Within ten minutes the machine was destroyed. Today, after twenty years of brain research, I still am unable to dissect the major regions of a rat's brain, a skill most of my students acquire during their first week.

Though he might break a few test tubes in an experiment, Axelrod remains the most skillful experimentalist I know. Far more crucial than manual dexterity is the ability to design the simplest possible experiment, one which can ask and answer the widest range of questions. Thus, in a typical experiment employing twenty test tubes, Axelrod might be addressing ten distinct, important issues. If only one pair of test tubes pays off, he will have made a new finding. By contrast, some scientists work for three or four years addressing a question that will provide major advances if the answer turns out to be yes but will lead to a dead end if the answer is no. This is a rather risky way to proceed. A better-designed experimental approach yields valuable insights regardless of whether the answer is yes or no. Not only should large-scale projects be addressed in this win-win manner, but every single experiment should be designed accordingly. Shots in the dark are permissible, but only if they can be handled in experiments that consume one or two days' effort, not months or years.

Nothing saddens me so much as witnessing brilliant young intellects addressing enormous efforts to trivial problems. Students should be taught how to select a productive area of inquiry. Often with new PhD students this is a hit-or-miss affair. In some laboratories, the first day that the student darkens the door, he is sent off by his lab director to the library with the admonition, "This is *your* PhD. Now it is your responsibility to select *your* research focus. Don't ask me. Spend the next several months in the library deciding what to do. Then return here and do it." The poor neophyte, with no prior research experience, is expected to exercise the most complex of scientific judgments—the selection of a research direction—on his own, with almost no advice from his mentor. Presumably these professors want their young charges to develop a capacity for making wise scientific choices. But what a silly way to do it! I have seen such students floundering for two or three years trying to solve inherently insoluble problems, most of which turn out to be fundamentally trivial.

Much of my time is devoted to devising new areas of potential research. I draw up long lists of potential projects and refine the lists week by week. When a new student joins the laboratory, we brainstorm together. If the student provides the entrancing and workable idea, we move forward with his or her project. If the best idea emanates from one of my own suggestions, then it wins out. I have no vested interest in one or another area. All that matters is that each student work on something of importance, with a reasonable probability of success.

I always try to achieve a good fit between a student and a project, but sometimes getting it right requires a certain amount of trial and error. Candace Pert was ill-suited for the acetylcholine project she initially undertook in my lab, but after she switched over to opiate-receptor research, everything seemed to fall into place for her. By contrast, opiate receptors were not Ian Creese's cup of tea. Though Ian made definitive contributions to this research, his true calling lay elsewhere, with the dopamine-receptor project.

Ian's PhD research had been conducted in a psychology department. He had tried to find out the types of behaviors that are regulated by the various dopamine neuronal systems, especially those dopamine systems in parts of the brain that were known to regulate emotional behavior. Ian wondered whether antischizophrenic drugs worked by blocking dopamine receptors in these structures. He developed elegantly selective means of destroying individual dopamine neuronal pathways without causing gross brain damage. Working in isolation in a small laboratory in Cambridge, he made more advances in this field than other people in dozens of high-powered laboratories throughout the world. But Ian could see the limitations of using only psychological methods for understanding brain function, and he applied to my laboratory in order to learn biochemical approaches.

In his first days in Baltimore, Ian was quite malleable about potential projects, and so I suggested that he explore the relative potencies of opiates in influencing intestinal contractions. Opiates are the best available drugs for diarrhea and, as we have seen, their influence on intestinal contractions can be readily measured in test-tube systems. In the same strips of intestinal tissue Ian could measure both the binding of opiates to receptor sites and contractile effects of the drugs. He found a remarkably close correlation between the relative potencies of a large number of opiates in influencing contractions of the intestine and in binding to receptors. This was the most definitive evidence that the opiate binding sites were indeed functional receptors, responsible for the pharmacological actions of opiates.

Though Ian had become deeply involved in the opiate work of the lab, he did not share Candace's overwhelming emotional "cathexis" to the field. Deep down inside, Ian was still a psychologist, interested in "higher" brain functions that might be relevant to major mental illness, especially schizophrenia. Thus, when he saw the results of Sam Enna's first experiments labeling dopamine receptors with radioactive dopamine, Ian was burning with curiosity and distressed that this was not to be his project. David Burt had just joined the laboratory and needed an area of focus, so, without much thought, I suggested that David inherit the dopamine-receptor work from Sam, as Sam was already overwhelmed with the GABA receptor. Up to that point Ian, with characteristic reticence, had not uttered a word to me about his desire to work on the dopamine receptor. When he could hold back no longer, he pleaded with me to let him join David on the project. I apologized for my obtuseness in not appreciating how Ian's earlier background would have made this so special an area for him. The unique combination of David and Ian resulted in a brilliantly productive body of work on dopamine receptors. Since graduating from my lab, Ian has continued with dopamine receptors and is now an acknowledged world leader in this burgeoning field.

As students gain experience in my lab, they become more independent and more often are the source of new ideas, which I try to encourage. Indeed, each day my discussions with students usually include one or more of the following utterances, "So—what happened? What do you think it means? Can you think of any really major consequence? Where shall we go from here? What experiment would you like to do next?"

Teaching a student how to design cleverly efficient experiments is more important than training students technical aspects of research. The design of research apparatus advances so rapidly that today I do not know how to operate half the machines in our laboratory. Computers now play a major role in all research, yet I haven't the vaguest notion of how to program one. Biochemists, including most of the students in my lab, can manipulate their experimental data with computers to generate elegant graphs with straight lines and curves of many motley shapes. I have witnessed young scientists plugging their data directly into the computer without ever glancing at the numbers that emerged from the individual test tubes. Frequently these same students also have not examined the test tubes themselves to see if the added chemicals were in solution, whether the solution was clear or cloudy, or whether a color change had occurred. They also have not looked carefully at their rats before sacrificing them. Did the drug treatment produce any change in the level of alertness, muscle strength, skin color?

By neglecting the immediacy of the crudest, raw data of direct observation, such scientists may miss the most dramatic findings. Breakthroughs stem frequently from chance observations. A scientist notices something peculiar and has the curiosity to suspect that something significant transpired. He or she then has the imagination to guess the relevance of the observation and to design simple, yet penetrating and answer-giving experiments. I encourage my students to heed what they see, hear, or smell with their experimental animals, tissues, and test tubes.

A recent experience in the lab illustrates this principle. John Wagner, a talented recent PhD student, had been interested in the role of calcium in heart function. He was able to measure a receptor for drugs that block calcium ion channels and are of considerable value in treating angina and high blood pressure. The receptor sites that John measured by the binding of the radioactive calcium antagonist drugs presumably are associated with calcium channels in membranes of the heart. John and I collaborated with cardiologists at Johns Hopkins in studies of a genetically determined heart disease in hamsters which resembles a serious form of cardiac hypertrophy in humans. We wondered whether the molecular cause of the disease might be an abnormality in numbers of these calcium channels. John discovered a doubling in the number of these channels in the hamsters with cardiac hypertrophy. The disorder appeared to be universal for all calcium channels in the body, not at all restricted to the heart. Indeed, John found the greatest increases in the brain.

Our cardiologist colleagues paid little heed to this biochemical aberration in brain tissue. However, as soon as John observed these changes in the brain he exploded with a "eureka" type of realization and came running to me: "Sol, I just realized that whenever I took those diseased hamsters out of the cage, they bit my hand. A normal hamster has never done that before. I think these hamsters have as big an emotional disease as they have a cardiac disease, and we may have our hands on the molecular cause!" I think John is absolutely correct, and we are exploring this question right now.

I have found that, as important as it is to train the intellect, a good mentor also needs to strengthen each student's soul. I don't try to play psychiatrist with my students. However, nurturing self-confidence is a crucial role for a mentor. Scientific training emphasizes analytical faculties. In graduate science courses, students read published papers and then vie with each other in efforts to attack the underlying rationale, the experimental techniques, and the conclusions. With so much focus on being critical and avoiding the pitfalls of poor experimental design, students often become overly self-critical. They knock down their own creative ideas so masochistically that little if any innovation survives. Such students never take chances, attempting only investigations whose results are assured of withstanding the stormiest criticism. Of course, if you can predict the consequences of your own experiments before they commence, your research is very likely to be *boring*. On several occasions, back when I was in Axelrod's lab, I felt that a body of work I had recently completed deserved to be burned, not published. He insisted that we review the findings, and when we did, he could find all sorts of fascinating implications in data that I had first felt to be useless.

The eminent psychologist Carl Rogers emphasized that "unconditional positive regard" by the therapist of his patient is the most healing element in the entire psychotherapeutic process. In a similar vein, the psychiatrist Jerome Frank views demoralization as the common thread among patients seeking psychotherapy. For Frank, the goal of therapy is a restoration of morale, a sense of high self-regard, an inward feeling of assurance, a strong ego. Perhaps it was my training as a psychiatrist, or perhaps it was Axelrod's influence, or maybe it was simply my experience as a parent—whatever the source, in the lab I strive to let my students know that they really matter to me, by praising their accomplishments and helping out when they encounter difficulties in their work. When personal problems intrude—and they do sooner or later for almost all graduate students during these high-pressure years—I try to be attentive. The payoff for these efforts is getting to witness firsthand one of the wonders of nature, a metamorphosis—the transformation of a young, tentative student into a self-possessed, probing, innovating intellect. Raising children, seeing patients, and nurturing students all have rewards in common when approached this way.

Like many other scientists, I often wonder at my good fortune, being paid a salary to pursue my favorite hobby. This sense of research as a joy is another ingredient of scientific discovery that I try to convey to my students—and, come to think of it, it's something else I learned from Julius Axelrod.

Notes

Acknowledgments

Index

Notes

1. *The Politics of Science*

1. Richard Severo, *New York Times*, October 23, 1969; *New York Times*, February 1, 1970.
2. Joseph Lelyveld, *New York Times*, January 12, 1970.
3. Alvin Shuster, *New York Times*, May 16, 1971.
4. James Watson, *The Double Helix* (New York: Atheneum, 1968; New American Library, 1969), p. 53.

2. *Drug Wars*

1. Quoted in D. I. Macht, "The history of opium and some of its preparations and alkaloids," *Journal of the American Medical Association* 64 (1915): 477–481.
2. Quoted in O. S. Ray, *Drugs, Society and Human Behavior* (St. Louis: Mosby, 1972), p. 180.
3. J. Jones, *The Mysteries of Opium Revealed* (London: Richard Smith, 1700), p. 32.
4. P. McCabe, "School days," *Rolling Stone*, February 18, 1971, p. 24. Quoted in Ray, *Drugs, Society and Human Behavior*, p. 180.

3. *Stalking the Opiate Receptor*

1. A. Goldstein, L. I. Lowney, and B. K. Pal, "Stereospecific and nonspecific interactions of the morphine congener levorphanol in subcellular fractions of mouse brain." *Proceedings of the National Academy of Science, USA* 68 (1971): 1742–1747.

4. Staking the Claim and Taking the Blame

1. J. D. Watson and F. H. C. Crick, "Molecular structure of nucleic acids: a structure for deoxynucleic acids," *Nature* 171 (1953): 737–738.
2. C. B. Pert and S. H. Snyder, "The opiate receptor: demonstration in nervous tissue," *Science* 179 (1973): 1011.
3. *Newsweek*, March 19, 1973, p. 55.
4. L. Terenius, "Stereospecific interaction between narcotic analgesics and a synaptic plasma membrane fraction of rat cerebral cortex," *Acta Pharmacologica et Toxicologica* 32 (1973): 317–320.

7. Designer Drugs

1. J. Hughes, H. Kosterlitz, and H. Morris, "Identification of two related pentapeptides from the brain with potent agonist activity," *Nature* 258 (1975): 577–580.

Acknowledgments

I would like to thank Drs. Alan Green, Jerome Jaffe, and William Bunney for providing background information about events in the opiate story. Dr. Seymour Kety and Professor John Dowling read early versions of the book and made useful comments, for which I am grateful. I would also like to acknowledge Laszlo Meszoly for the elegant illustrations that help to clarify the science, and Susan Wallace for her magnificent editing of the text.

Index